Field Theory of Nonimaging Optics

This book aims to overcome the traditional ray paradigm and provide an analytical paradigm for Nonimaging Optics based on Field Theory. As a second objective, the authors address the connections between this Field Theory of Nonimaging Optics and other radiative transfer theories.

The book introduces the Field Theory of Nonimaging Optics as a new analytical paradigm, not statistical, to analyze problems in the frame of nonimaging geometrical optics, with a formulation based on field theory of irradiance vector $\vec{D}$. This new paradigm provides new principles and tools in the optical system design methods, complementary to flowline method, overcoming the classical ray paradigm. This new Field paradigm can be considered as a generalization of the ray paradigm, and new accurate and faster computation algorithms will be developed. In a parallel way, the advance in the knowledge of the principles of Field Theory of Nonimaging Optics has produced clear advances in the connection between nonimaging optics and other apparently disconnected theories of radiation transfer. The irradiance vector $\vec{D}$. can be considered as the macroscopic average of Poynting vector, with a clear connection with radiation pressure. Lorentz geometry techniques can also be applied to study irradiance vector $\vec{D}$. There are clear thermodynamic connections between the nonimaging concentrator and Stefan-Boltzmann law of radiation. From this thermodynamic connection, nonimaging optics and irradiance vector $\vec{D}$. can also be studied from a phase space point of view.

This book is intended for researchers, graduate students, academics and professionals looking to analyze, design and optimize optical systems.

Field Theory of Nonimaging Optics

Ángel García Botella,
Roland Winston, and Lun Jiang

CRC Press
Taylor & Francis Group
Boca Raton London New York

CRC Press is an imprint of the
Taylor & Francis Group, an **Informa** business

First edition published 2024
by CRC Press
6000 Broken Sound Parkway NW, Suite 300, Boca Raton, FL 33487-2742

and by CRC Press
4 Park Square, Milton Park, Abingdon, Oxon, OX14 4RN

CRC Press is an imprint of Taylor & Francis Group, LLC

ISBN: 978-0-367-54344-0 (hbk)
ISBN: 978-0-367-55163-6 (pbk)
ISBN: 978-0-367-55160-5 (ebk)

DOI: 10.1201/9780367551605

Typeset in CMR10 font
by KnowledgeWorks Global Ltd.

Publisher's note: This book has been prepared from camera-ready copy provided by the authors.

Contents

1 The Light Field and the Flowline Design Method   **1**

 1.1 Introduction 1

 1.2 Sphere Ellipse Paradox 3

 1.3 The Irradiance Vector 4

 1.4 Irradiance Vector, Geometrical Vector Flux and Étendue . . 6

 1.5 The Edge Ray Principle and the Cone of Edge Rays 9

 1.6 Basic Properties of a Light Field $\vec{D}$ 12

 1.7 Surface Integral of $\vec{D}$ 13

 1.8 The Line Integral of $\vec{D}$ 15

 1.9 Flowline Design Method 16

 1.10 The Role of Coordinate Systems 18

 1.11 Field of Non-Lambertian Sources 19

2 Flowline Method in Nonimaging Designs   **23**

 2.1 Compound Parabolic Concentrator CPC 23

 2.2 Hyperparabolic Concentrator HPC 27

 2.3 Compound Elliptical Concentrator 33

 2.4 Coordinate Systems in the Flowline Design Method 35

 2.5 3D Hyperboloid 38

 2.6 One-Sheeted Hyperboloid 40

 2.7 Flowline Asymmetric Nonimaging Concentrating Optics . . . 45

 2.8 Ideal Source-Receiver Transmission Design 47

3 Field Theory Elements   **53**

 3.1 Classification of Nonimaging Optics Fields 53

 3.2 Geometrical Description of Modulus of $\vec{D}$ 54

 3.3 Contour Integrals of the Light Field 57

 3.4 $\vec{D}$ Produced by an Arbitrary Plane Source 61

 3.5 Vector Potential and Gauge Invariance 63

 3.6 Some Irradiance Pattern Computation Examples 70

 3.7 The Curl of $\vec{D}$ and Quasipotential Fields 77

 3.8 Basic Introduction to Lorentz Geometry 82

 3.9 Application of Lorentz Geometry to the Evaluation of $\vec{D}$ . . 85

4 The Irradiance Vector in Optical Media   **93**
   4.1 $\vec{D}$ Vector at Interface between Refractive Media   93
   4.2 Orthogonal Refractive Interfaces   96
   4.3 Refracted Cone of Edge Rays   101
   4.4 $\vec{D}$ through Refractive Media   106
   4.5 $\vec{D}$ through Reflective Media   111
   4.6 Lorentz Formalism to Compute $\vec{D}$ through Refractive Media   114
   4.7 Using Lorentz Formalism to Compute $\vec{D}$ through Reflective
       Media   119
   4.8 $\vec{D}$ through Inhomogeneous Refractive Medium, Curved Cones
       of Edge Rays   121

5 Thermodynamic Basis of the Irradiance Vector   **125**
   5.1 Introduction   125
   5.2 Relations between $\vec{D}$ and Thermodynamic Variables   125
   5.3 Thermodynamic Origin of Nonimaging Optics   127
   5.4 Blackbody Radiation in a Cavity and the Radiation Pressure   132
   5.5 Kirchhoff Radiation Law   134

6 Phase Space in Nonimaging Optics   **139**
   6.1 Introduction to Phase Space in Nonimaging Optics   139
   6.2 Irradiance Vector $\vec{D}$ in Phase Space   152
   6.3 The $\frac{\theta_i}{\theta_o}$ Concentrator   154
   6.4 Cone Concentrator as Ideal 3D Phase Space Transfer Device   155
   6.5 Fermi Proof of Phase Space Volume or Étendue Conservation   156

A The Edge Ray Theorem   **159**
   A.1 Introduction   159
   A.2 The Continuous Case   159
   A.3 The Sequential Surface   165
   A.4 The Flowline Mirror Case   166

Bibliography   **169**

Index   **175**

1

The Light Field and the Flowline Design Method

1.1 Introduction

Geometrical optics deals with rays and is usually identified with the tracing of such rays through optical systems, as in the design of lenses. The most accepted definition of a ray is the trajectory of radiation energy from a point in the source to a point in the detector. The trajectories of rays are governed by Fermat's principle. Following the idea of radiation energy of rays, we can consider the definition of radiometry as the measurement of energy of radiation and the determination of how this energy is transferred from a source, through a medium, to a detector. Traditionally radiometry assumes that the propagation of radiation energy can be treated using the laws of ray propagation. The term photometry is limited to the ability of radiation to produce a visual sensation in the human visual system. This subject is of considerable importance, dealing as it does with such applications as nonimaging optics, photography, color science and astronomy. One of the first treatises on photometry was written in 1760 by Johan Heinrich Lambert [1]. Returning to the important aspect of ray definition, that it is defined for a point in the source, this means that to study real optical systems with extended light sources, it is needed to analyze a great number of rays (typically tens of millions). In this sense Mehmke in 1898 [2] suggested the definition of irradiance vector $\vec{D}$ produced by an extended radiation source in any point in space, which units are $\frac{W}{m^2}$ and with direct relation to scalar irradiance. This idea was developed mainly by Fock, Gershun and Moon [3], [4], [5]. Gershun [4] also suggested the relation between irradiance vector $\vec{D}$ and radiation pressure force. More recently Winston and Welford [6] [7], the fathers of what today we know as nonimaging optics, developed the so-called flowline design method, using the field lines produced by such a vector to design ideal concentrators. Flowline design method was a crucial step forward in the development of the field theory of nonimaging optics. Following those developments, in this century new advances in the field theory of nonimaging optics have been made [8], [9]. These backgrounds suggest the use of field theory concepts in the analysis of the propagation of radiation, employing as a basic analysis object the vector $\vec{D}$. Field theory is one of the most powerful methods to study

DOI: 10.1201/9780367551605-1

physical phenomena. The first successful field application was accomplished by Euler (1736) in his work in hydrodynamics. Advances in field theory were made by Laplace (1799), Fourier (1822) and Maxwell (1861), and finally at the beginning of last century Einstein [10] developed the celebrated relativistic field theory. Today field theory is applied to nearly all of physics. On the other hand, radiative transfer combines principles of geometrical optics and thermodynamics to characterize the flow of radiant energy at scales large compared with its wavelengths and with time intervals large compared to its frequency. Central to the theory of radiative transfer are the principles of radiometry, the measurement of light. It is interesting to note that Chandrasekhar in his book "Radiative transfer" [11] remarks that net radiation flux behaves like a vector, but he does not pursue this point further. Basic thermodynamics laws, like the Stefan-Boltzmann law and Kirchhoff radiation law, talk about radiative transfer. In chapter 5 we will connect these fundamental laws with irradiance vector $\vec{D}$.

The objective of this book is to provide an overview of the field theory of nonimaging optics as a powerful optical design technique, extending the flowline design method. In fact, there are no fundamental drawbacks to extending the field theory of nonimaging optics to the design of imaging optics, but it has not been developed yet. We will develop theoretical tools and apply them to practical examples to analyze the transfer of radiation energy in optical systems with real extended light sources, using classical field theory elements, moving beyond the traditional paradigm of raytracing. It is important for the reader to note that throughout the book we will continually use the two main properties of vector $\vec{D}$:

1. The components of $\vec{D}$ are proportional to the scalar irradiance at any point in space.
2. The field lines of $\vec{D}$ provide the geometry for ideal optical devices.

These two properties of $\vec{D}$ allow us to compare the results obtained by the application of the field theory concepts with raytracing simulations, which provides a fast and robust quasi-experimental validation of the results. This book is organized as follow: In chapter 1 we study the basic field concepts, irradiance vector and flowline method; chapter 2 shows well-known example of optical elements designed using flowline technique and field theory concepts; in chapter 3 we introduce field theory concepts to analyze free propagation of radiation, including vector potential and Lorentz geometry; in chapter 4 we study optical systems with refractive and reflective elements, including interface factors, quasipotential interface and Lorentz geometry applied to refractive and reflective media; chapter 5 shows the connection between basic thermodynamics variables and laws with irradiance vector $\vec{D}$; and finally in chapter 6 we study the phase space as the connecting element between field theory and thermodynamics of radiation transfer.

1.2   Sphere Ellipse Paradox

Because imaging optics treats the source and its destination (sink or detectors) of light rays as points, it processes the sources and sinks by simplifying them. In other words, the analysis of the optical system is based on point-to-point mapping through rays. Although such an underlying assumption works very well, it fails to address the thermodynamic rules that the optical sources or sinks with extended surfaces are required to observe, and this can lead to paradoxes. One classical example [12] is the paradox of an elliptical spherical chamber, as shown in fig. 1.1, where the assumption that the radiation source and sink can be simple points will create a dilemma that violates the second law of thermodynamics. Consider a spherical-ellipsoid chamber with perfect reflectivity for its inner walls. The point object A is positioned at the center of a spherical reflecting cavity, and it is also at the focus of an elliptical reflecting cavity; the point object B is at the other focus. If we start A and B at the same temperature, the probability of radiation from B reaching A is clearly higher than A reaching B. The light rays directly from A and the reflected rays in the spherical part of the cavity will be redirected to A, besides the light rays from B and reflected in the elliptical part of the cavity will be redirected to A. One can easily validate such a result with ray tracing. So we conclude that A warms up while B cools off, despite starting at the same temperature, resulting in a violation of the second law of thermodynamics (heat only flows from higher temperature to lower temperature). The paradox is resolved by making A and B extended objects, no matter how small. In fact, a physical object with temperature has many internal degrees of freedom and cannot be point-like. Obviously in such an example the geometrical principle of rays being reflected from one focus to the other focus for ellipses, and from the central point of a spherical mirror and back for spheres, is not at fault. Rather, it is the assumption that we can treat optical objects as simple points, producing a result that does not exist in reality. Simple points can neither produce rays nor receive rays in radiation heat transfer. Their usage can help to conceptualize and analyze an optical system. However, in many cases, the over-simplification of point-like sources and sinks for radiation heat transfer produces self-conflicting results, such as those of the sphere-ellipse paradox. The solution is not to fully give up on ray tracing. Instead, in addition to using rays and points to help guide the optical designing process, we must also consider thermodynamics as one of the fundamental principles that cannot be violated through such a designing process. Field theory of nonimaging optics at its core addresses such a challenge.

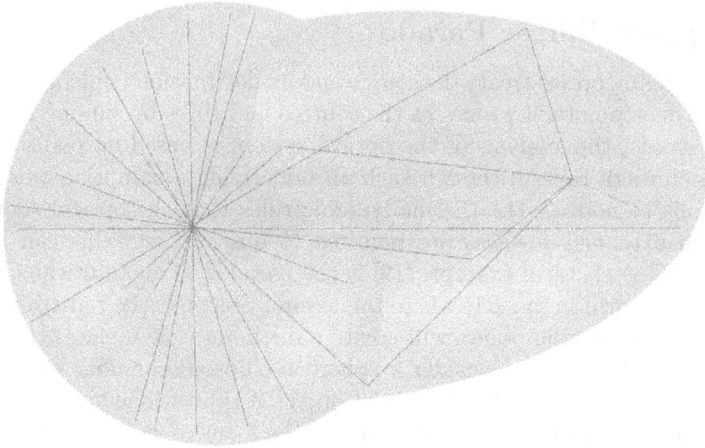

FIGURE 1.1: Spherical-elliptic paradox for point sources

1.3   The Irradiance Vector

Conventional photometry and radiometry deal with scalar physical magnitudes, brighness, flux, irradiance. The idea of considering irradiance as a vector was proposed by Mehmke in 1898 [2] and partially developed by Fock [3], Gershun [4] and Moon [5]. The main advantage of this approach is that we can define a physical vector field associated with $\vec{D}$ vector at any point in space and then apply mathematical analysis and conservation laws well established in vector field theory instead of conventional scalar analysis or statistical raytracing methods. This vector image of light propagation provides a new point of view of geometrical optics and overcomes the raytracing paradigm, mainly due to the fact that raytracing is based on point sources, which are not real sources, as we have stated in previous section. We will see that the irradiance vector is well defined for a real extended source, and in the limit case of a small source (point source) the field lines of the irradiance vector $\vec{D}$ converge to rays.

Scalar irradiance E is a magnitude defined upon a surface, for example: irradiance on the earth's surface, on a detector, on a table; in consequence it has a cosine dependence on the incident angle, being ϕ the angle between the direction of incidence of radiation and the normal to the surface where we measure E

$$E = E_{max} \cos(\phi) \tag{1.1}$$

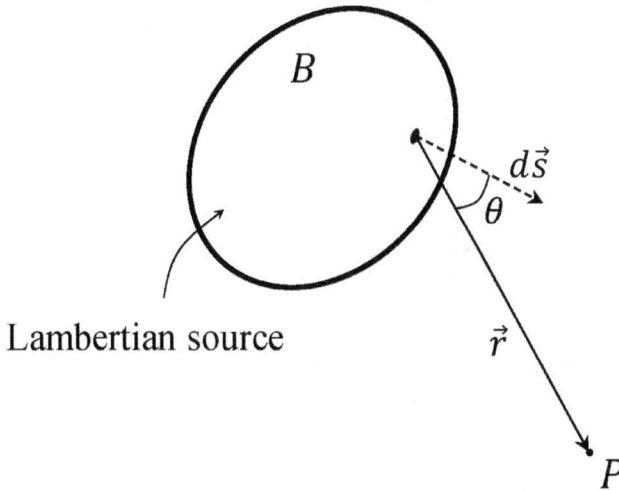

FIGURE 1.2: Basic geometry to obtain $\vec{D}$ at point P from a Lambertian source of brightness B

This equation clearly points us to the existence of a vector $\vec{D}$ defined in any point in space which accomplishes the basic scalar product:

$$E = \vec{D} \cdot \vec{N} \tag{1.2}$$

Gershun [4] provides a complete proof of the existence of the irradiance vector, which in fact is analogous to the proof used in the analytic theory of heat. A precise definition of $\vec{D}$ is that at any point P of the light field there is a vector which is independent of the choice of coordinate system, and which possesses the property that its projection upon any direction is numerically equal to the difference of scalar irradiance of the two sides of a plane element placed at P and normal to that plane.

For a extended Lambertian source Fig. 1.2, $\vec{D}$ at point P can be computed with the surface integral [5][6]:

$$\vec{D} = \int_s \frac{B\vec{r} \cdot d\vec{s}}{r^3} \vec{u}_r = \int_s \frac{B\cos(\theta)ds}{r^2} \vec{u}_r \tag{1.3}$$

where B is the brightness of the source, $\frac{\cos(\theta)ds}{r^2}$ is the solid angle subtended by the infinitesimal element of area ds to the point P and $\vec{u}_r$ is the unit vector in the direction of $\vec{r}$; the integral domain is the surface source. As a consequence of the existence of that vector we can automatically define its associated field vector as the region of the space where it is defined. The relation of that vector with scalar irradiance is that the D_z component is the scalar irradiance incident at point P upon the XY surface, and the radiometric units of $\vec{D}$ are $\frac{W}{m^2}$. The irradiance vector $\vec{D}$ can also be interpreted as the field vector for radiation pressure force [13]. The relation between the irradiance vector and radiation pressure force to a particle is

$$\vec{F}_{rad} = \kappa \vec{D} \tag{1.4}$$

where κ is a scalar magnitude which depends on the area of the particle, the absorption or reflection properties of the particle and the speed of light c. $\vec{F}_{rad}$ is the radiation pressure force. The connection between the irradiance vector $\vec{D}$ and electromagnetic theory can be done through the Poynting vector [14]. Another fundamental connection of the irradiance vector arises from the Stefan-Boltzman law, which relates the scalar emittance of the surface of a blackbody also expresed in $\frac{W}{m^2}$ to temperature in the form

$$E = \sigma T^4. \tag{1.5}$$

Both the magnitudes of emittance from a surface and irradiance upon a surface are connected by the irradiance vector $\vec{D}$. An interesting conncetion between eq. 1.5 and geometrical optics arises from the concentration limit. Considering the sun as a blackbody radiator at 6000 K, the radiation flow at the solar surface is about $58,6\,W/m^2$, the solar constant is about $1,35\,mW/m^2$ and both magnitudes are related by the theoretical radiation concentration limit obtained from geometrical considerations

$$C_{max} = \frac{1}{sin^2\theta_s} \sim 44000 \tag{1.6}$$

where θ_s is the angle of the cone subtended by the solar surface at a point on earth, about 0.27°.

1.4   Irradiance Vector, Geometrical Vector Flux and Étendue

As we have mentioned in the introduction of this chapter, the development of flowline design method was a crucial step forward in the development of a field theory for nonimaging optics. Winston and Welford developed it [6] [7] using the concept of geometrical vector flux $\vec{J}$, which is a purely geometrical vector based on the concept of étendue. In this section we are going to establish the relation between the irradiance vector $\vec{D}$ and its homologous geometrical vector flux $\vec{J}$. Prior to doing this, we must introduce the concept of étendue, a basic optical concept for the analysis of real extended sources. For an optical extended source the étendue is the product of the area of the source and the solid angle of the radiation emitted by the source; for an optical system the étendue is the product of the area of the entrance pupil and the solid angle subtended from the pupil. It is well known that étendue is conserved in a loss free optical system, and in section 6.5 we offer Fermi's proof of the conservation of étendue [15]. The differential element of étendue dU is given by

$$dU = dp_x dp_y dx dy \qquad (1.7)$$

where p_x and p_y are the optical direction cosines, which means the ray direction cosines multiplied by the refractive index n. The total étendue entering an optical system is

$$U = \int dp_x dp_y dx dy \qquad (1.8)$$

and emerges unchanged if there is no attenuation and if the system is such that no rays are stopped or reflected back. This result was used by Winston [16] to find the maximum concentration which could theoretically be attained by a concentrator. The differential element dU is a measure of the number of rays entering the surface element $dx dy$ within the solid angle $dp_x dp_y$. The optical momentum $\vec{p} = n(\cos\theta_x, \cos\theta_y, \cos\theta_z)$ can be expressed in spherical coordinates fig. 1.3 as

$$\vec{p} = n(\sin\theta_z \cos\psi, \sin\theta_z \sin\psi, \cos\theta_z) \qquad (1.9)$$

being the infinitesimal element

$$dp_x dp_y = \left(\frac{\partial p_x}{\partial \theta_z} \frac{\partial p_y}{\partial \psi} - \frac{\partial p_x}{\partial \psi} \frac{\partial p_y}{\partial \theta_z} \right) = n^2 \cos\theta_z \sin\theta_z d\theta_z d\psi = n^2 \cos\theta_z d\Omega \qquad (1.10)$$

where $d\Omega$ is the infinitesimal solid angle. Then for the radiation from a Lambertian source

$$dU = n^2 \cos\theta_z d\Omega dx dy \qquad (1.11)$$

where θ_z is the angle between the pencil beam and the z axis. Equation 1.11 should be written in the form

$$dU = \cos\theta_z dJ dx dy \qquad (1.12)$$

where dJ is the number of rays in the elementary bundle. By simple comparison between eq. 1.8 and eq. 1.12, we can write for the z component of vector $\vec{J}$

$$J_z = \int dp_x dp_y. \qquad (1.13)$$

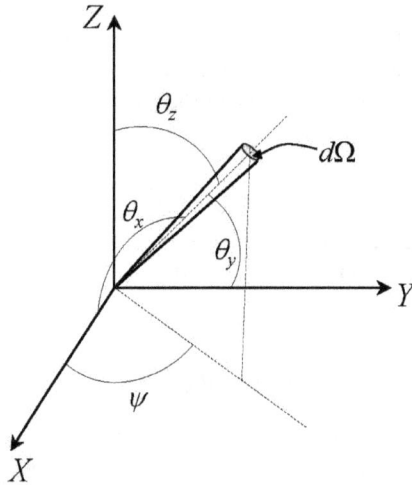

FIGURE 1.3: Solid angle for a infinitesimal beam

We can similarly define

$$J_x = \int dp_z dp_y,$$
(1.14)

and

$$J_y = \int dp_z dp_x.$$
(1.15)

We can consider these quantities to be the components of a vector $\vec{J}$; we shall call it geometrical vector flux. From this definition of $\vec{J}$, it is easy to see that the dimensions of $\vec{J}$ is the stereoradian (sr), considering that for a Lambertian source all rays have the same brightness B measured in $\frac{W}{srm^2}$. This provides the simple relation between the geometrical vector flux $\vec{J}$ and the irradiance vector $\vec{D}$

$$\vec{D} = B\vec{J}.$$
(1.16)

It is also possible to obtain eq. 1.3 from $\vec{D}$ using solid angle considerations. Then in conclusion, $\vec{J}$ is well defined for Lambertian sources with constant brightness , while the irradiance vector $\vec{D}$ is defined for any kind of source. In section 1.11 we will study $\vec{D}$ produced by Gaussian beams, and all other chapters and sections of this book will be concerned with Lambertian sources,

No edge rays   Source

S_1 S_2

Edge rays

Edge rays

Optical element

R_1 R_2
Detector

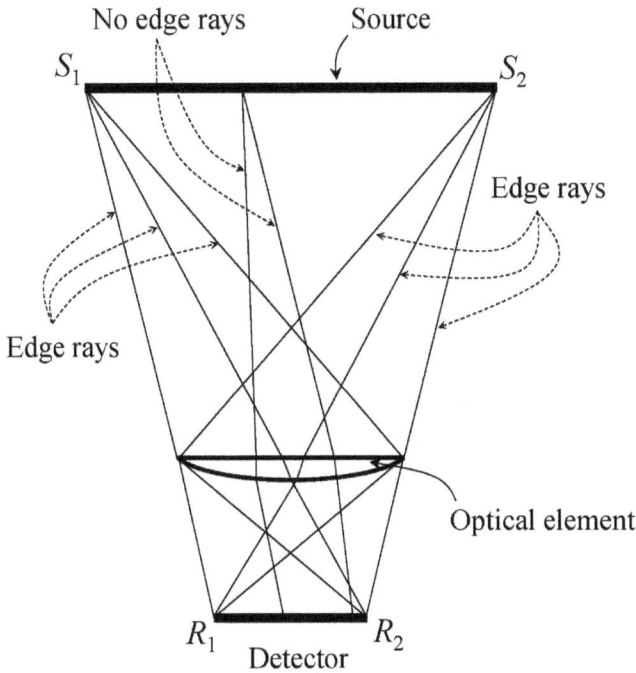

FIGURE 1.4: Illustration of the edge ray principle

for which the irradiance vector $\vec{D}$ and geometrical vector flux $\vec{J}$ are completely analogous.

1.5   The Edge Ray Principle and the Cone of Edge Rays

The Edge Ray Principle and the Cone of Edge Rays Some forms of the edge ray principle have been used to design nonimaging concentrators. The edge ray principle states that nonimaging devices can be designed by the mapping of edge rays from the source to the edge of the target [17]. This guarantees that all rays inside the source will be transferred inside the target. Figure 1.4 illustrates the edge ray principle. For a source $S_1 S_2$, a lens and a target $R_1 R_2$, all rays which emerge from S between S_1 and S_2 will arrive at the target between R_1 and R_2. We provide a detailed demonstration of this principle in appendix A.

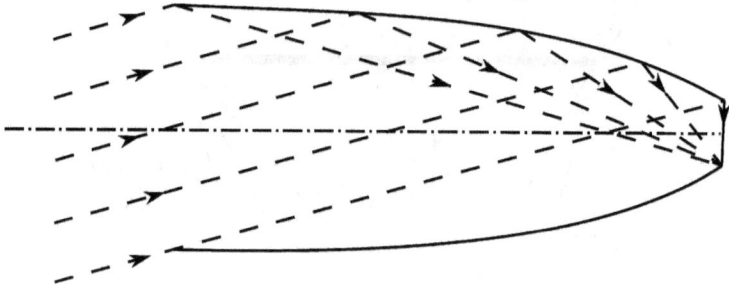

FIGURE 1.5: Edge rays incident on a CPC

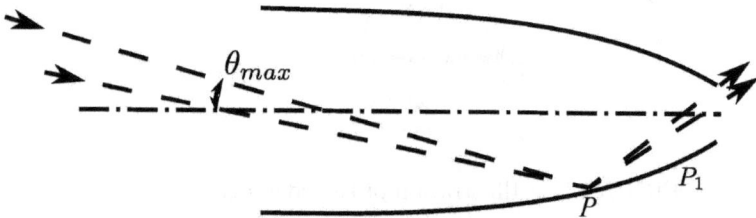

FIGURE 1.6: The inner rays will be redirected to the target

In this section we study some examples of nonimaging devices which can be well defined using the edge ray theorem and some consequences of it. The most celebrated example is the Compound Parabolic Concentrator (CPC) [18]. Figure 1.5 shows that the CPC maps source edge rays, rays incident at an angle θ_{max}, to the target edge. Figure 1.6 also shows that all rays incident at an angle lower than θ_{max} will be redirected to the target. From the edge ray principle it is possible to define the important concept of a cone of edge rays. Let there be a Lambertian source and a point in space, P. The rays from the contour of the source to the point P form the cone of edge rays, fig. 1.7. We will show later that using Stoke's Theorem it is possible to compute the irradiance vector $\vec{D}$ from an integral over this cone of edge rays. The base of the cone can have any shape—circular, square, or polygonal; the important property is that for a Lambertian source, this cone of edge rays provides all the information needed to obtain $\vec{D}$, and of course this cone is different at any point in space, and also different if we put the point P inside a refractive optical component or after a reflexive surface. Another interesting property of the cone of edge rays is that as we move away from the Lambertian source, the cone narrows, so that far away from the source, at the limit of a point source, the cone of edge rays becomes infinitesimal, and we can say that a ray is an infinitesimal cone of edge rays, as shown in fig. 1.8.

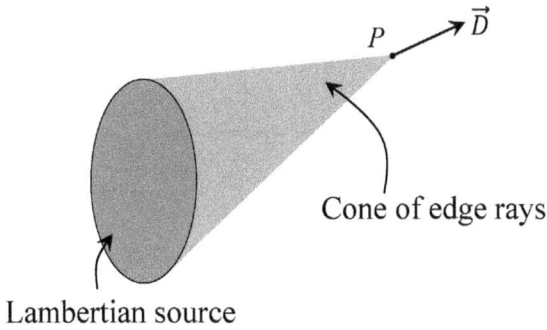

P $\vec{D}$

Cone of edge rays

Lambertian source

FIGURE 1.7: Cone of edge rays

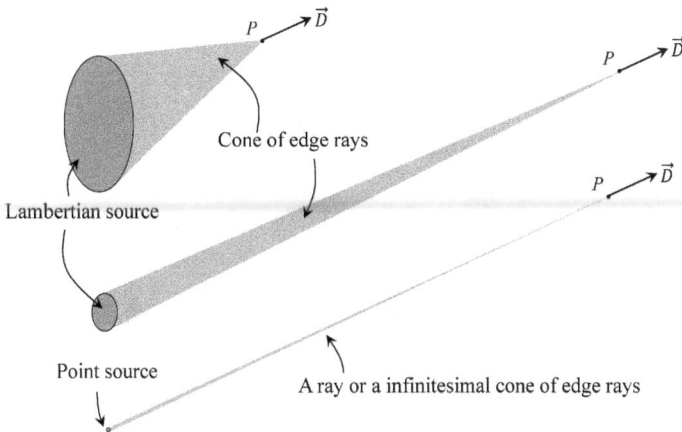

P $\vec{D}$

P $\vec{D}$

Cone of edge rays

P $\vec{D}$

Lambertian source

Point source

A ray or a infinitesimal cone of edge rays

FIGURE 1.8: Far away from the light source the cone of edge rays becomes a ray

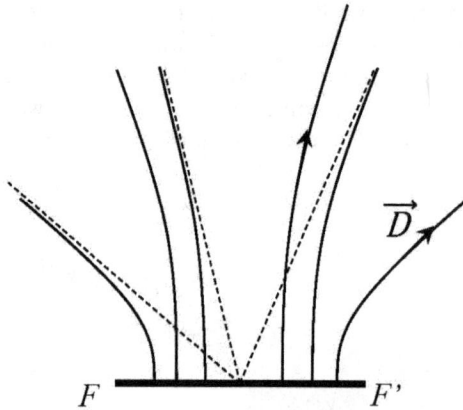

FIGURE 1.9: Field lines of $\vec{D}$ produced by a strip uniform source

1.6   Basic Properties of a Light Field $\vec{D}$

A basic issue to consider in the analysis of a light field is the superposition: as a light field is clearly a vector field the superposition of different light sources will produce a vector superposition of

$$\vec{D}_{tot} = \sum_i \vec{D}_i. \qquad (1.17)$$

As a result, the pressure radiation force upon any particle in the field will also be a superposition of vector forces. A different situation appears if we want to compute scalar irradiance: it depends on a surface, and the effect of the source can be blocked by the detector surface. For the computation of scalar irradiance produced by several sources on a surface, it is necessary to take into account the orientation of the surface with respect to the sources, using eq. 1.2; this effect is known as view factor [19].

As a basic property of the irradiance vector $\vec{D}$, the field line distribution produced by basic light source geometries is interesting to study; remember that field lines are lines tangent to the vector field at any point in space. Gershun [4] studied several basic geometries. The field produced by spherical source is radial. For a uniform luminous strip, the field lines are confocal hyperbolas with foci at F and F' at the edges of the strip; this example was studied by Gershun, Moon and Winston. Figure 1.9 shows the field line pattern. This simple field line pattern has two important properties which will be useful in our knowledge of the field. First the asymptotic behavior of the hyperbolas: looking at this field line pattern far away from the source, we can observe that the hyperbolas converge to straight lines. This means that in the limit case of a point source, the field lines converge to form rays. Note that rays are the trajectory of the radiation energy. This basic property allow us to consider the field theory of nonimaging optics as a generalization of ray theory to extended sources. The second basic property of the field pattern is that each hyperbola has a constant difference in optical path length from the focus to any point in the hyperbola, which is the basic geometrical property of the hyperbola; this property will take an important role in identifying each field line and defining the vector potential function. Another interesiting example is the field lines produced by a uniform luminous circular disc, which are again represented by confocal hyperbolas with foci at F and F'; by rotating these hyperbolas about the axis of the disc we obtain hyperboloids of revolution, which forms flux tubes. In chapter 2 we will study analytic equations for $\vec{D}$ produced by a circular disk. As another example, Fock [3] studied the field distribution produced by an elliptic disc with semiaxis α and β; he used the vector potential concept to study it, and he concluded that field lines are orthogonal to the ellipsoids of equation

$$\frac{x^2}{\lambda} + \frac{y^2}{\lambda - \beta^2} + \frac{z^2}{\lambda - \alpha^2} = 1 \qquad (1.18)$$

where λ is the parameter of the family of ellipsoids; one-sheeted hyperboloids are orthogonal to this ellipsoids. In chapter 2, we will return to the study of interesting properties of one-sheeted hyperboloids as concentrators. Figure 1.10 shows a one-sheeted hyperboloid as a flux tube produced by an elliptic disc.

More recently Moon [5] applied the irradiance vector to illumination engineering; he obtained a general expression for the irradiance vector produced by polygonal sources of uniform illumination. The irradiance vector and also irradiance tensor have been applied to rendering [20], and the irradiance tensor has applications in scattering and in non-isotropic sources or interfaces.

1.7 Surface Integral of $\vec{D}$

The net radiative flux through a surface can be computed from the irradiance vector $\vec{D}$ as a surface integral, using the definition of flux concept:

$$F = \int_A \vec{D} \cdot d\vec{A}. \qquad (1.19)$$

Hyperbolic meridional section

Elliptic cross section

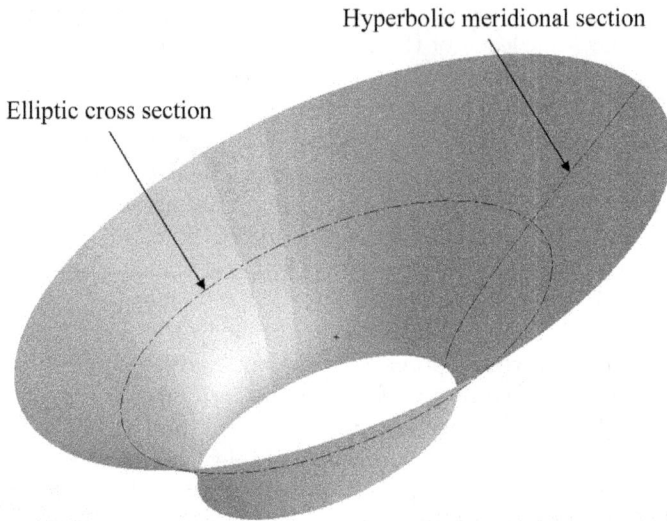

FIGURE 1.10: One-sheeted hyperboloid, flux tube of an elliptic disk source

The flux through a closed surface is related to the divergence of $\vec{D}$ by Gauss's theorem

$$\oint_A \vec{D} \cdot d\vec{A} = \int_v \operatorname{div} \vec{D} dv. \tag{1.20}$$

Gershun [4] showed a complete demonstration that for all the points of the light field at which there is no emission or absorption of light:

$$div\vec{D} = 0. \tag{1.21}$$

Winston [19] arrived at the same conclusion which is an expression of conservation of energy principle. An interesting element in the classical field theory is the concept of a flux tube formed by an infinite number of field lines, fig. 1.11, and with an input and an output surface. Flux tubes have interesting properties: for instance, the total flux F is conserved through the flux tube and provides an alternative way to understand the important concept of the étendue conservation theorem in the propagation of light and its relation with the irradiance vector $\vec{D}$ [7]. The total étendue entering a system, as we have seen in section 1.4, is

$$U = \int \int dp_x dp_y dx dy \tag{1.22}$$

where p_x and p_y are direction cosines in rectangular coordinate systems. The total étendue entering an optical system emerges unchanged if there is no

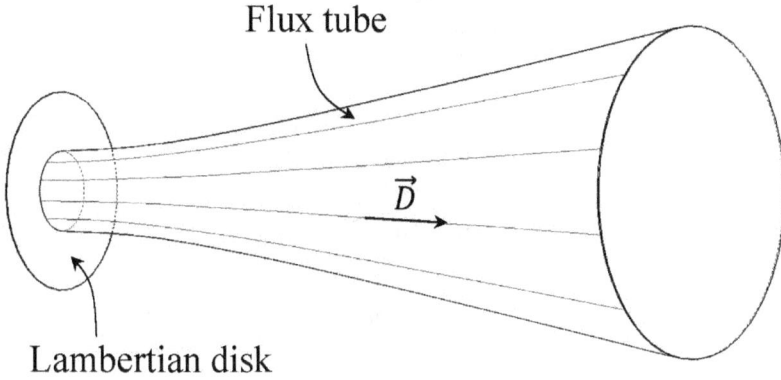

FIGURE 1.11: Flux tube

source or sink for rays. For such a system the total flux also emerges unchanged. For rectangular coordinate systems, the total flux in eq. 1.19 can be rewritten as the flux emerging through a surface orthogonal to z axis:

$$F_s = \int_s D_z dx dy. \tag{1.23}$$

By comparison between eq. 1.22 and eq. 1.23 for the ideal flux tube produced by a Lambertian source of fig. 1.11, we can obtain

$$D_z = B \int dp_x dp_y \tag{1.24}$$

and similarly,

$$D_x = B \int dp_y dp_z, \quad D_y = B \int dp_x dp_z \tag{1.25}$$

where B is the brightness of the source, in agreement with eq. 1.16.

1.8   The Line Integral of $\vec{D}$

In the frame of vector field theory, the line integral of the vector field through a closed line is related to the existence or not of the scalar potential. Only for simple light sources, like spherical source, can we say that a scalar potential exists, but we cannot obtain a general expression for the scalar potential of $\vec{D}$. Nevertheless we can make some analysis of the line integral, which can be useful. First we can use the Stoke's theorem for the closed line integral of $\vec{D}$ in free space :

$$\oint_l \vec{D} \cdot \vec{dl} = \int_s \nabla \times \vec{D} \cdot \vec{ds}. \tag{1.26}$$

As a simple example, we can study the $\nabla \times \vec{D}$ for a spherical source with brightness B; the $\vec{D}$ for this case is

$$\vec{D} = \frac{B\vec{u}_r}{r^2} \tag{1.27}$$

using the property

$$\nabla \times (\psi\vec{A}) = \psi(\nabla \times \vec{A}) + (\nabla\psi \times \vec{A}). \tag{1.28}$$

Considering the brightness as a scalar function, we can write

$$\nabla \times \vec{D} = B\,\nabla \times \frac{\vec{u}_r}{r^2} + \nabla B \times \frac{\vec{u}_r}{r^2}. \tag{1.29}$$

The first term to the right side of eq. 1.29 is zero. Then

$$\nabla \times \vec{D} = \frac{\nabla B \times \vec{u}_r}{r^2}, \tag{1.30}$$

for general $\nabla \times \vec{D}$ is not zero and the field of a spherical source has no scalar potential; only the particular case of a uniformly bright spherical source has scalar potential. We will return to this question in chapter 3.

1.9   Flowline Design Method

From those basic concept in field theory, in the last decades of the past century Winston developed the so-called flowline design method, which was the first design technique based on a field theory of nonimaging optics [16]. The flowline design method provides ideal concentrators and was one of the powerful techniques in the advent of what today we know as nonimaging optics. The method is based in placing a perfect mirror on the surface of a flux tube; then the field lines local to the mirror side will be undisturbed. To see that, note that at point P on the mirror, the normal component of $\vec{D}$ is zero. Now this zero normal component is made up of equal and opposite components

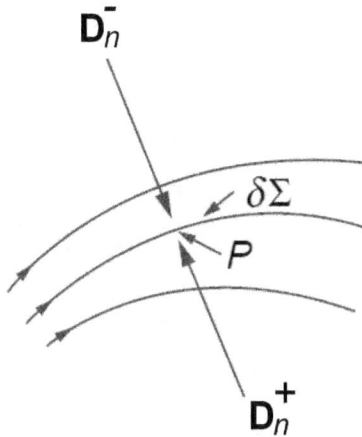

FIGURE 1.12: Detailed balance of rays

$\vec{D}_n^+ = -\vec{D}_n^-$ from rays coming from either side of the mirror, as shown in fig. 1.12. But the effect of inserting the mirror is to remove $\vec{D}_n^-$ and replace it with $\vec{D}_n^+$ with the reverse sign, according to the law of reflection, so there is no local disturbance.

We have noted that a mirror surface placed along field lines of ithe irradiance vector produced by a source will not disturb the flow pattern under a certain condition. This condition is that the zero normal component of $\vec{D}$ must result from the fact that, for each ray r_1 incident on one side of the surface, there must be a corresponding ray r_2 incident on the other side such that r_2 is in the direction of the reflection of r_1; we call this a condition of detailed balance of rays. Since light rays, and therefore field lines of $\vec{D}$, are reversible, we see that a way to construct a concentrator with maximum theoretical concentration is to place mirrors in the field lines under conditions of detailed balance of rays. This method is what has been called the flowline method. It can be also expressed as placing perfect mirrors along flowlines, which do not disturb the optical field, fig. 1.13. This constitutes an entirely new perspective on optical design. In fig. 1.13 we have built mirrors with the shape of cones, which are the geometry of flux tubes produced by a spherical source, the orange in our example. Another example of the flowline design method is the hyperboloid of revolution as a concentrator. As we have mentioned, the field lines produced by a Lambertian strip are confocal hyperbolas with focal points at the edge of the strip. In chapter 2, we study and analyze in detail the flowline design method and show some examples of an ideal reflector design.

FIGURE 1.13: Perfect mirrors with the geometry of field lines do not disturb the optical field

1.10   The Role of Coordinate Systems

Field theories in physics provide integral or differential equations as techniques to obtain required physical magnitudes, scalars, vector or tensors. A key point in analytically solving those integral or differential equations is to find an appropriate coordinate system that fits the geometry of the particular configuration. The key point is to find an appropriate coordinate system that allows us to apply the separation of variables method to solve the integral or differential equations. It is also possible to solve the field problem if a uniform source conforms in shape to the coordinate surface of a known coordinate system. As a simple example, we can study the field produced by a uniform spherical radiation source (like the sun, or the light reflected by the orange of fig. 1.13) using a spherical coordinate system. The surfaces orthogonal to the sphere of $R = $ constant can form a cone, which consists of the field lines of $\vec{D}$, as shown in fig. 1.14. An analogous procedure can be done with other orthogonal coordinate systems. In chapter 2 we will use oblate spheroidal coordinates (η, ϕ, ψ) to analyze the field $\vec{D}$ produced by a uniform circular disk source; we also use ellipsoidal coordinates (η, ϕ, λ) to analyze the field produced by a uniform elliptic disk source. Toroidal coordinates will be used to transport the radiation from a Lambertian circular disk to any point or orientation in space. In fact, it is possible to generalize and state that each orthogonal coordinate system produces ideal optical devices by fitting the uniform source

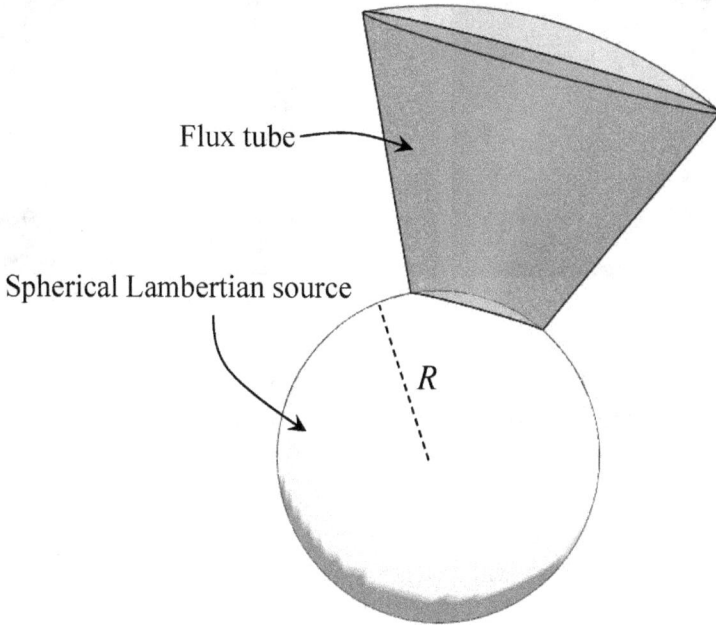

FIGURE 1.14: Spherical coordinates, cones as flux tubes of spherical sources

to the shape of a constant coordinate surface, so that orthogonal coordinates provide the direction of $\vec{D}$ and, by flowline design method, can provide a ideal optical device.

In [21], Moon tabulates equations for 40 coordinate systems, but using other techniques like conformal mapping, [22] an infinitude of orthogonal coordinate systems is possible.

1.11 Field of Non-Lambertian Sources

Equation 1.3 provides a basic law to compute $\vec{D}$. Throughout the book we will study properties of $\vec{D}$ and practical examples in which we will use Lambertian light sources. This means that the brightness B in eq. 1.3 is constant and we can get it from the integral. In fact the Lambertian sources are affected by the surface factor $\cos(\theta)$, but we have already considered this geometric factor in

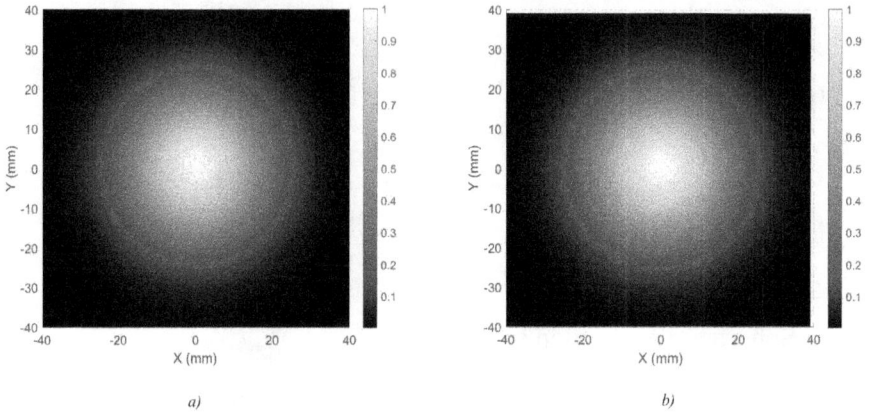

FIGURE 1.15: Comparison of scalar irradiance at $z = 100\,mm$ plane for a Gaussian source with $w = 0.13$, a) irradiance vector $\vec{D}$ computation, b) raytracing computation

eq. 1.3. In this section we will analyze $\vec{D}$ produced by Gaussian light sources in an introductory way, providing the basic methodology for studying these sources. We will consider angular Gaussian beams in way that for any source any point in the source emits a Gaussian brightness light distribution in the form

$$B = B_0 e^{-\frac{\theta^2}{w}} \tag{1.31}$$

where B_0 is the brightness for $\theta = 0$ and w controls the width of the Gaussian beam. Then the $\vec{D}$ can be computed as the surface integral, also referred to in fig. 1.2:

$$\vec{D} = \int_s \frac{B_0 e^{-\frac{\theta^2}{w}} \cos(\theta)\vec{r}}{r^3} ds \tag{1.32}$$

In fact, it is possible to use the same procedure to compute $\vec{D}$ for anisotropic Gaussian beams by defining two angles, θ_x and θ_y, one for each axis, and taking care in the integration process. As an example, fig. 1.15 shows two irradiance distributions at a detector plane, for both Gaussian beams, a) computed from the integral of eq. 1.32 and b) computed by raytracing simulation, for a square source of 10 mm side, with the detector placed 100 mm from the source plane and $w = 0.13$. The maximum irradiance has been normalized to one.

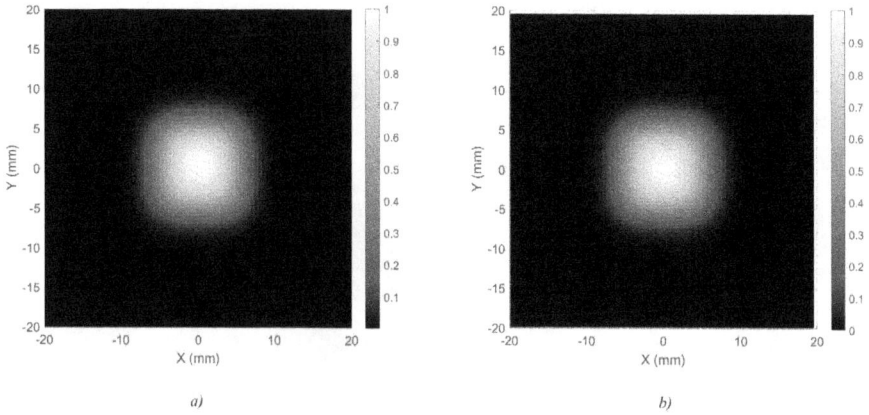

FIGURE 1.16: Comparison of scalar irradiance at $z = 15.5\,mm$ plane for a Gaussian source with $w = 0.13$, *a*) irradiance vector $\vec{D}$ computation, *b*) raytracing computation

Figure 1.16 shows the same comparison between irradiance distributions, but in this case the detector placed is 15.5 mm from the source.

Then as a result, each light beam distribution will produce its own vector field pattern with its own properties. Nevertheless, in this book we will study vector fields produced by Lambertian sources. Note that most of the standard light sources, including LED, can be modeled using a Lambertian surface with any geometry and adding one or more optical components.

2

Flowline Method in Nonimaging Designs

As we have mentioned, the flowline design method was a crucial step forward in the comprehension of $\vec{D}$: It provides a simple and direct tool to design perfect optical devices, clearly overcoming the raytracing paradigm and showing the intimate relation between $\vec{D}$ and radiation energy transfer. It is focused on reflective optical design, and it uses the conservation of radiation flux through flux tubes of a vector field produced by a radiation source. Placing perfect mirrors at the surfaces defined by those flux tubes will obtain perfect reflective elements, which can be used as ideal concentrators. Note that the flowline method is focused on the direction of $\vec{D}$; the modulus of $\vec{D}$ does not matter for this chapter. Winston [16] developed this method and applied it to studying several new concentrators, including the celebrated Compound Parabolic Concentrator (CPC), also known as the Winston cone, leading to what today we know as nonimaging optics.

In chapter 1, we have shown that the field lines of $\vec{D}$ produced by a Lambertian disk source are confocal hyperbola, with the focus at the edge of the disk. Of course different geometries of the source will produce different geometries of the field lines, and then those news geometries can provide new ideal concentrators by using the flowline design method, placing perfect mirrors with the geometry of those field lines and flux tubes. In this chapter we are going to study several examples of these geometries, which have interesting applications as nonimaging optics devices. The first examples will be obtained by the analysis of a 2D field and then by building nonimaging devices by rotation or extrusion of those field lines. The last examples in the chapter are 3D devices without rotational or translation symmetry.

2.1 Compound Parabolic Concentrator CPC

As a first example, we are going to analyze the well-known Compound Parabolic Concentrator (CPC)[23], but before that, we are going to consider $\vec{D}$ produced by 2D Lambertian sources.

Consider a Lambertian strip source AA' in the x, y plane, and then consider the point P in the plane fig. 2.1. The brightness of all the rays arriving at P from the strip is equal. Then by direct integration $\vec{D}$ points along the bisector

DOI: 10.1201/9780367551605-2

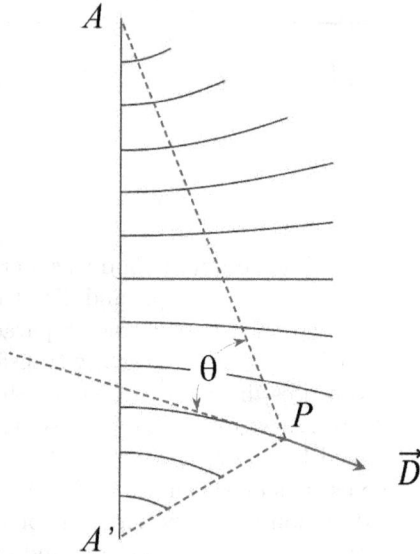

FIGURE 2.1: Field lines of a 2D strip AA'

of the angle APA'. This condition is well known; it is satisfied by a hyperbola with focal points at A and A', and thus the field lines produced by a 2D strip form a family of confocal hyperbola. The modulus of $\vec{D}$ at P can be obtained by a line integral from A to A' of eq. 1.3, in such a way that $|\vec{D}| = 2\sin\theta$ where the angle $APA' = 2\theta$. Notice that the difference of distances from P to the focal points $|AP| - |A'P|$ is constant along a field line. This magnitude plays a relevant role, which we will return to in chapter 3. On the other hand, it is easy to see that the field lines produced by a semi-infinite strip are confocal parabolas with the focal point at the end A of the strip. By moving the focal point A' to infinity the confocal hyperbolas become confocal parabolas.

Following this procedure we can consider the field lines produced by a infinite wedge, fig. 2.2. For this configuration we can divide the 2D space in three regions; in regions where only one face of the wedge is seen, the field line pattern is the same as that for a semi-infinite strip, conformal parabolas. From any point where both faces are seen, the effect is of an infinite extended Lambertian strip, and the field line patterns are parallel straight lines as indicated in fig. 2.2.

The next geometry to consider, which produces the CPC, is the truncated wedge of fig. 2.3 $RQQ'R'$, which is infinitely extended back from the apex with an included angle $2\theta_0 = 60°$. We have to consider four different regions

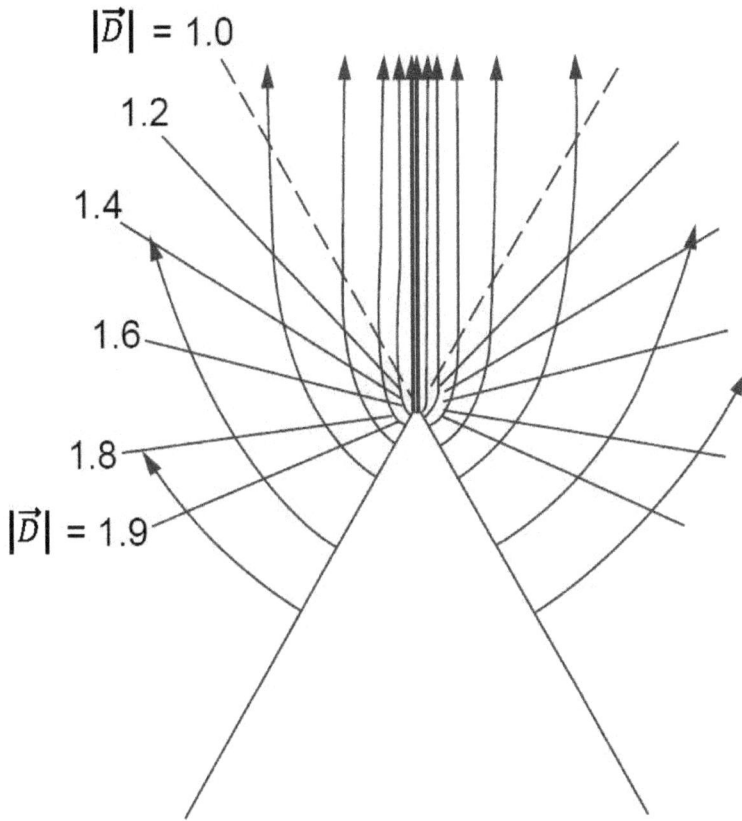

FIGURE 2.2: Field lines produced by a Lambertian source with the shape of a wedge

in calculating the vector $\vec{D}$. These are separated by the boundary lines of the wedge produced, as shown in dashed lines, and they are labeled regions A (and A'), B, C (and C') and D. A point P in region A receives light only from face RQ, so the distribution of $\vec{D}$ is the same as for a semi-infinite strip: that is, the field lines are confocal parabolas with focal point at Q. If point P is in region B it receives light only from the strip QQ', and again we have the result that the field lines are confocal hyperbolas with Q and Q' as focal points. If point P is anywhere in region D the field lines are straight lines orthogonal to QQ'. Finally, in region C the point P receives radiation from RQ and QQ'; the direction of $\vec{D}$ thus bisects the angle between a constant direction (parallel to QR) and the radius of fixed point Q', and thus the field lines in region C are confocal parabolas with a focal point at Q' and axis through Q' and parallel to QR.

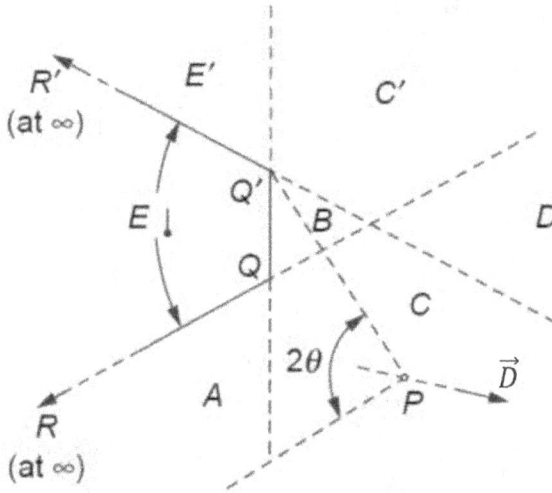

FIGURE 2.3: A truncated wedge of angle 2θ

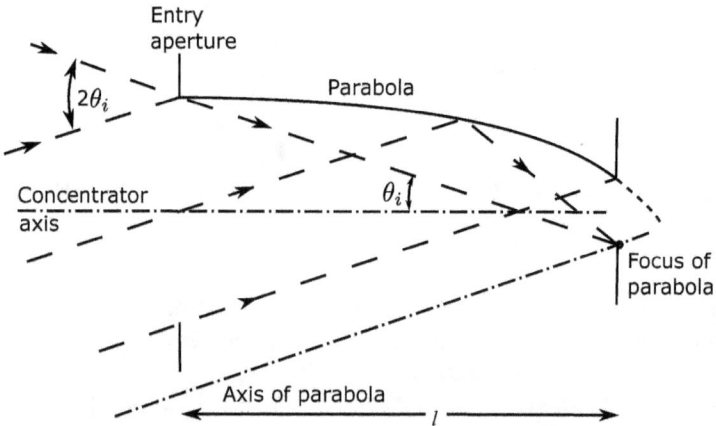

FIGURE 2.4: Construction of the CPC profile

Figure 2.3 shows that the field line which starts at the vertex of the wedge produces the well-known Compound Parabolic Concentrator. The geometry is composed of a tilted parabola with its focal point in the opposite edge of the exit aperture, fig. 2.4. From a ray optics point of view, all the rays incident on the entry aperture with an angle lower than the tilted angle θ_0 will emerge at the exit aperture; this produces a theoretical ideal concentrator. The 3D device can be obtained by rotation of the field line profile. The CPC was the first nonimaging optics device and has great interest as a solar concentrator device. Its ability to collect the radiation with an incident angle lower than θ allows the device to work for solar applications without the need of expensive and complex solar tracking systems. Figure 2.5 shows the transmission angle curve for a CPC with an acceptance angle of 16°, computed by ray tracing, and approaching very close to an ideal square shape.

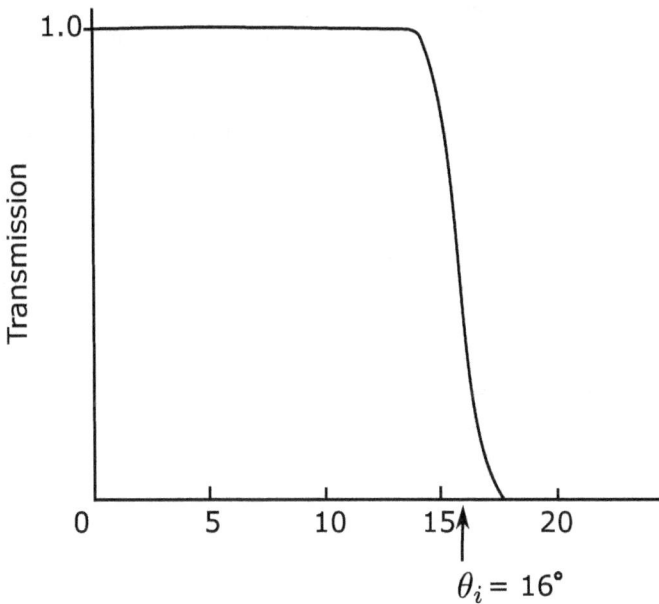

FIGURE 2.5: Transmission angle curve for a CPC with the acceptance angle $\theta_i = 16°$

2.2 Hyperparabolic Concentrator HPC

The flowline method has been used to study the three-dimensional (3D) compound parabolic concentrator (CPC) [16] as a component originating from the light field produced by a Lambertian emitter in the form of a 2D truncated wedge, but a 2D truncated wedge can produce other new concentrators by considering multiple reflections [8]. Many developments and modifications of the CPC have been studied [24], including a dielectric-filled CPC, a truncated CPC, and a two-stage CPC. Shatz and Bortz studied a global optimization procedure [24] to obtain reflective concentrators having superior performance to that of the 3D CPC. Now we focus our interest on the study of the field from a Lambertian emitter in the form of a two-dimensional (2D) truncated wedge, and we show that the field lines of this source can define a new higher-order family of concentrators, the HPCs, which improve on the performance of the 3D CPC, which is the lower limit of performance of the 3D HPC, and have the thermodynamic limit of concentration as the theoretical upper limit of performance .

Considering the light field produced by a truncated wedge, we can study the field lines that start at the upper part of the wedge, but not at the vertex. See fig. 2.6. These field lines produce new concentrators, which are HPCs, with the same acceptance angle as a CPC but with a smaller exit aperture diameter. These new concentrators have the shape of a hyperbola continuously joined with a tilted parabola, with its focus at the vertex of the wedge. Figure 2.6 shows that the profile has an inflection point, which proves that its order is higher than 2. The union between these two conics must fulfill three conditions:

1. The focus of the hyperbola, F_o, and the focus of the parabola, F_p, must coincide.

2. The union point U is the intersection point between the hyperbola and a line that passes through the nearest focus of the hyperbola, F, and cuts the axis of the hyperbola at an angle θ.

3. The axis of the parabola, which passes through its focus F_p, cuts the axis of the hyperbola at an angle θ.

These conditions ensure a continuous union, with a continuous slope, between hyperbola and parabola.

The starting and the ending points of the profile have 0 slope. Figure 2.7 shows a sketch of this profile. With this geometry, in the limiting case when the focal length of the hyperbola is the radius of the exit aperture, $f = a'$, the HPC becomes the CPC. By its geometric design, it is easy to prove that the 2D HPC is an ideal concentrator. All the rays incident on the entry aperture of a 2D HPC at angles minor or equal to θ will be redirected, by the parabolic section of the concentrator, to the segment $F' - F$, and their second and successive reflections will cause them to emerge from the exit aperture.

The construction of HPCs can be done by rotational symmetry (3D HPC) and by translational symmetry (2D HPC). Figure 2.8 shows three 3D HPCs with different values for the hyperbola focal length f and the same value for acceptance angle θ and exit aperture radius a'.

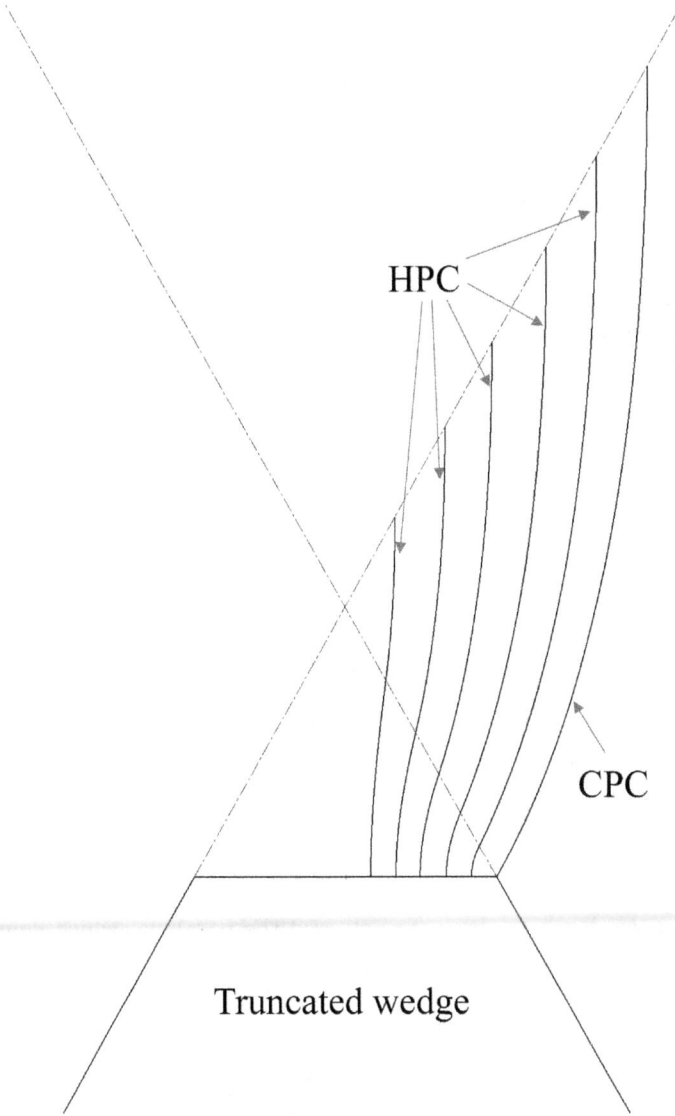

FIGURE 2.6: Field lines of $\vec{D}$ for the 2D truncated wedge, it shows HPC and CPC profiles

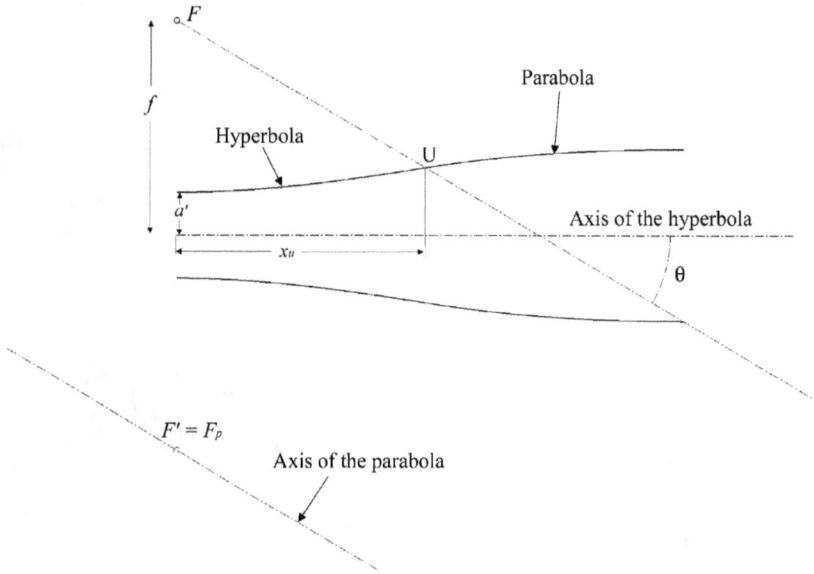

FIGURE 2.7: Geometric profile of the HPC

To obtain an analytic characterization, the HPC must be geometrically defined by three parameters, unlike the CPC, which needs only two parameters to be defined. The parameters of the HPC are f, the focal length of the hyperbola; θ, the acceptance angle; and a', the radius of the exit aperture. These parameters completely determine the shape and the overall length L of the HPC (fig. 2.7).

$$L = \left(f + \frac{a'}{\sin(\theta)} \right) \cot(\theta) \qquad (2.1)$$

The union point between the hyperbola and the parabola, which defines the length of the hyperbolic part of the concentrator x_u (fig. 2.7), can be calculated by the intersection between the hyperbola and the line that passes through the focus. Using the HPC parameters, it is

$$x_u = \frac{(f^2 - a'^2)(f \tan(\theta) - a' \sec(\theta))}{f^2 \tan^2(\theta) - a'^2 \sec^2(\theta)}. \qquad (2.2)$$

This point is the inflection point in the profile of the concentrator. By this construction, the parabola has a focal length of

$$f_p = \left(f + \frac{a'}{\sin \theta} \right) \sin \theta = f \sin \theta + a'. \qquad (2.3)$$

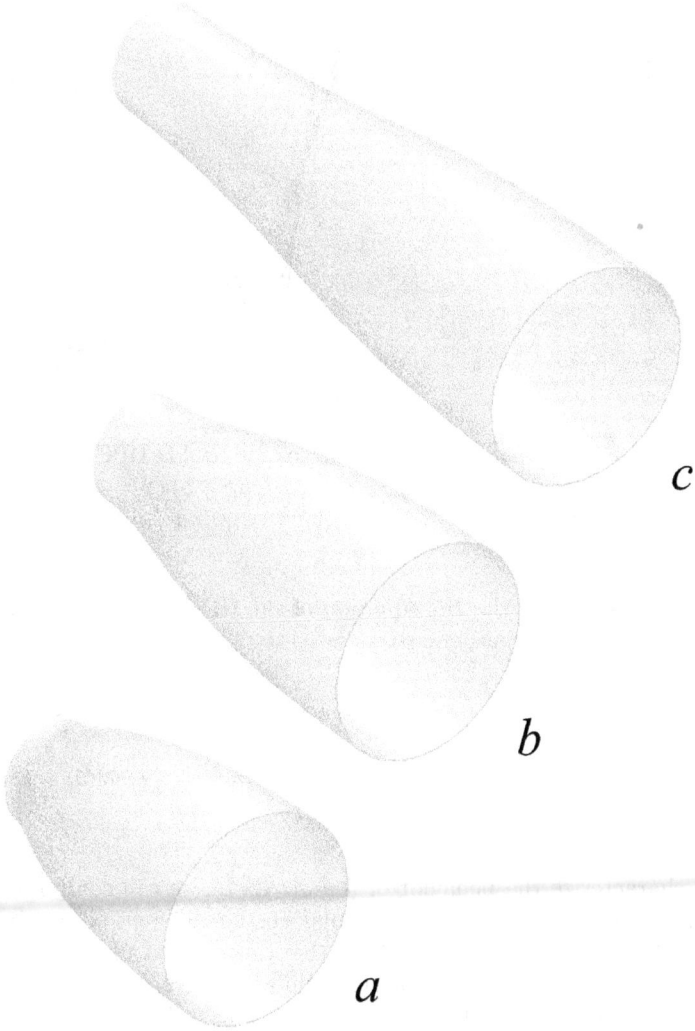

FIGURE 2.8: Three 3D HPCs with different hyperbola focal lengths of $f_a = 18\,mm$, $f_b = 30\,mm$, and $f_c = 60\,mm$, and the same acceptance angle $\theta = 30°$ and radius of exit aperture $a_o = 12\,mm$.

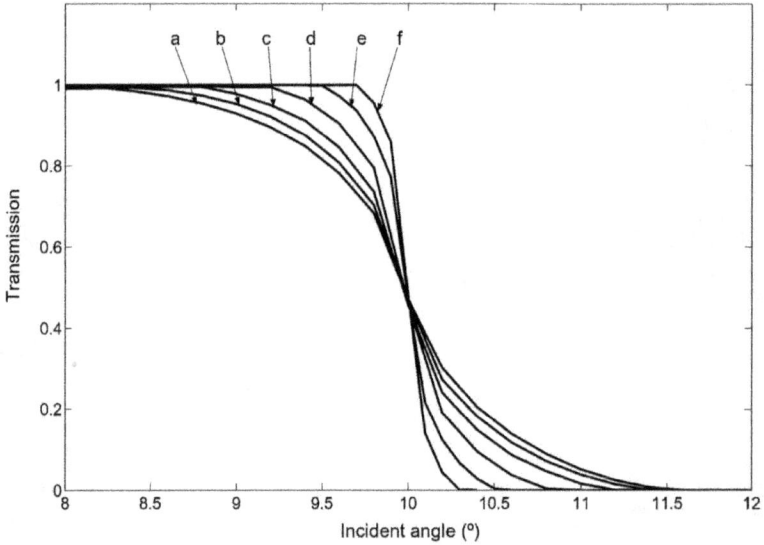

FIGURE 2.9: Transmission-angle curves for six 3D HPCs with $\theta = 10°$, $a_0 = 12\,mm$: CPC (curve a), $f = 18\,mm$ (curve b), $f = 30\,mm$ (curve c), $f = 60\,mm$ (curve d), $f = 120\,mm$ (curve e), and $f = 240\,mm$ (curve f)

It is also possible to write the equation of the HPC as an implicit function of its parameters, the hyperbolic part and tilted parabolic part:

$$\frac{y^2}{a'^2} - \frac{x^2}{f^2 - a'^2} = 1, \;\; 0 < x < x_u \tag{2.4}$$

$$((y + f)\cos\theta + x\sin\theta)^2 = 4(f\sin(\theta) + a')(x\cos\theta - (y + f)\sin(\theta) + f\sin\theta + a')$$
$$\text{for } x_u < x < L$$

$$(2.5)$$

To characterize the behavior of the 3D HPC, we can compute the transmission properties by ray tracing simulations for different parameter configurations. We analyze the dependence on the transmission angle curve versus the focal length of the hyperbola for three different acceptance angles, $\theta = 10°$, $\theta = 30°$ and $\theta = 50°$. The dependence on the radius of the exit aperture a0 is basically a scale factor. The way to build scaled HPCs with the same acceptance angle θ and the same transmission-angle curve is to multiply, by a scale factor of N, not only the a_0 parameter, as in the CPC, but also the f parameter. Figure 2.9 plots transmission-angle curves for six 3D HPCs with an acceptance angle $\theta = 10°$, radius $a_0 = 12\,mm$, and reflectance $\rho = 1$; these curves correspond to a CPC (curve a), an HPC with $f = 18\,mm$ (curve b), an HPC with $f = 30\,mm$ (curve c), an HPC with $f = 60\,mm$ (curve d), an HPC with $f = 120\,mm$ (curve e), and an HPC with $f = 240\,mm$ (curve f). It shows

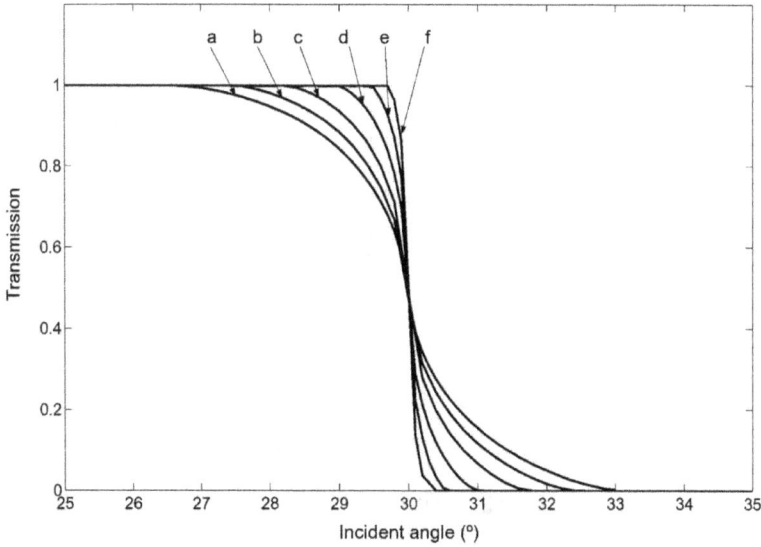

FIGURE 2.10: Transmission-angle curves for six 3D HPCs with $\theta = 30°$, $a_0 = 12\,mm$: CPC (curve a), $f = 18\,mm$ (curve b), $f = 30\,mm$ (curve c), $f = 60\,mm$ (curve d), $f = 120\,mm$ (curve e), and $f = 240\,mm$ (curve f)

that the CPC transmission-angle curve becomes the lower limit of performance of the HPCs, and that the total transmission of the concentrators increases with f. At the theoretical limit of $\lim f \to \infty$, the 3D HPC behaves as an ideal infinite source concentrator, with a steep transmission-angle curve, thus achieving the thermodynamic limit of concentration. Similar transmission-angle curves have been computed for HPCs with acceptance angles of $\theta = 30°$ (fig. 2.10) and $\theta = 50°$ (fig. 2.11).

2.3 Compound Elliptical Concentrator

In the previous sections we have assumed the light source to be at infinity for applications for solar energy concentration, CPC and HPC. There are, however, obviously cases where we would like to transport radiation from a source at a finite distances, and for this purpose the Compound Elliptical Concentrator (CEC) was designed [25].

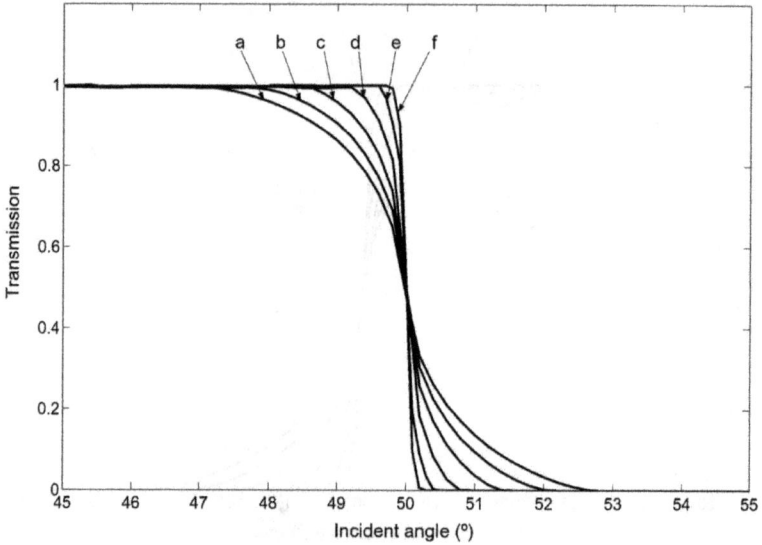

FIGURE 2.11: Transmission-angle curves for six 3D HPCs with $\theta = 50°$, $a_0 = 12\,mm$: CPC (curve a), $f = 18\,mm$ (curve b), $f = 30\,mm$ (curve c), $f = 60\,mm$ (curve d), $f = 120\,mm$ (curve e), and $f = 240\,mm$ (curve f)

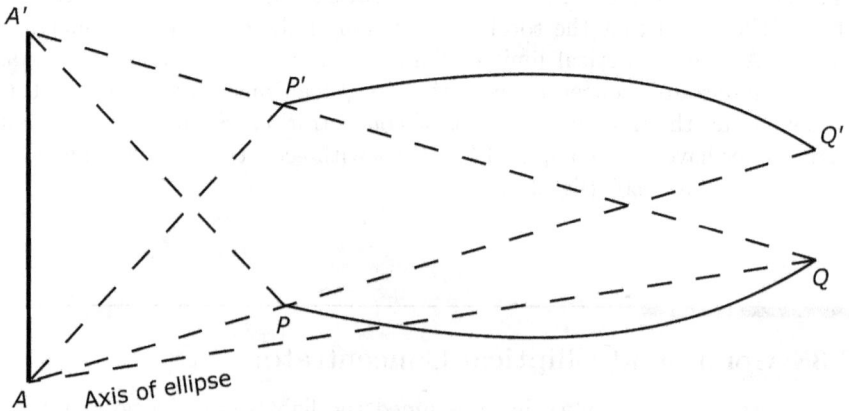

FIGURE 2.12: A CEC concentrator for the object AA' at a finite distance

In fig. 2.12 let AA' be the finite source and let QQ' be the desired position of the absorber. The reflecting surface QP' has the shape of an ellipse with foci at A and Q. For a 3D system the complete surface would be obtained by rotating this ellipse about the axis of symmetry. We can show that as a

2D system this has the maximum theoretical concentration by noting that all rays from AA' that enter the device do emerge. We have from the fundamental properties of ellipses that the sum of the distances from the two foci to any point of the curve is a constant, that is

$$AP + PQ' + QQ' = AP' + PQ, \text{ or } AP' - AP = Q'Q, \qquad (2.6)$$

so that the étendue measured at the output end is $2Q'Q$. Since QQ' is perpendicular to the axis, this must mean that rays emerge from all points of the exit aperture with their direction cosines distributed uniformly over ± 1; that is, this system has the maximum theoretical concentration ratio. For a 3D case with rotational symmetry a straightforward calculation gives for the étendue [25]

$$\frac{\pi^2(AP' - AP)^2}{4}. \qquad (2.7)$$

Nevertheless, this system for 3D will turn back some radiation in an analogous way to CPC. In fact, it is possible to build HyperElliptical Concentrators (HEC) in an analogous way to HPC to reduce the radiation that turns back. Then analogously at the limit where the focal length of the hyperbola approaches infinity, the turn back radiation approaches 0. Barnett [26] analyzed the $\vec{D}$ produced by CEC and Greenman [27] studied source configuration to obtain field lines of $\vec{D}$ to produce CEC. In the next section we will study field lines with the geometry of CEC in the frame of orthogonal coordinate systems.

2.4 Coordinate Systems in the Flowline Design Method

As we have mentioned in section 1.10, a key point in solving one physical field problem is finding an appropriate coordinate system that fits the geometry of the problem. In fact the optimum way to solve a particular field theory problem is to develop a coordinate system that fits the particular geometry. As we mentioned, there are some standard coordinate systems [21], and there are also techniques to build orthogonal coordinate systems to solve field problems. Particularly interesting is the technique of conformal mapping [28].

In previous sections we have analyzed several devices, concentrators, which are developed from basic properties of conic sections. All of them can also be obtained from the analysis of particular orthogonal coordinate systems. As the first example we will consider the CPC and also HPC : both can be obtained from the same orthogonal coordinate system obtained using the 2D truncated wedge as a geometric boundary. Figure 2.13 shows the 2D orthogonal coordinate system produced by the truncated wedge, divided into four sections

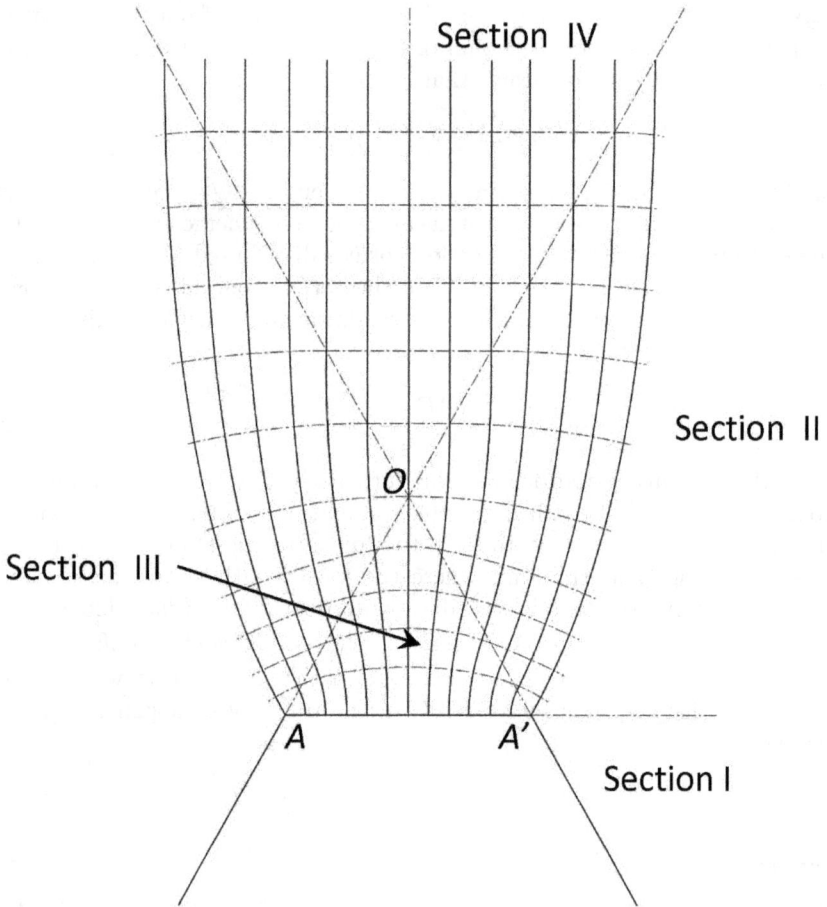

FIGURE 2.13: Coordinate system produced by a truncated wedge

separated by dashed straight lines. In region I the coordinates are defined by confocal parabolas with the focal point at A' and with the axis along the line $A'O$; in region II the coordinates are defined by confocal parabolas with the focal point located at A and with the axis parallel to the line $A'O$ and passing through the point A; in region III the coordinates are defined by confocal hyperbolas with focal points at A and A' (solid lines) and confocal ellipses with focal points at A and A' (dashed lines); and finally region IV is composed by straight lines.

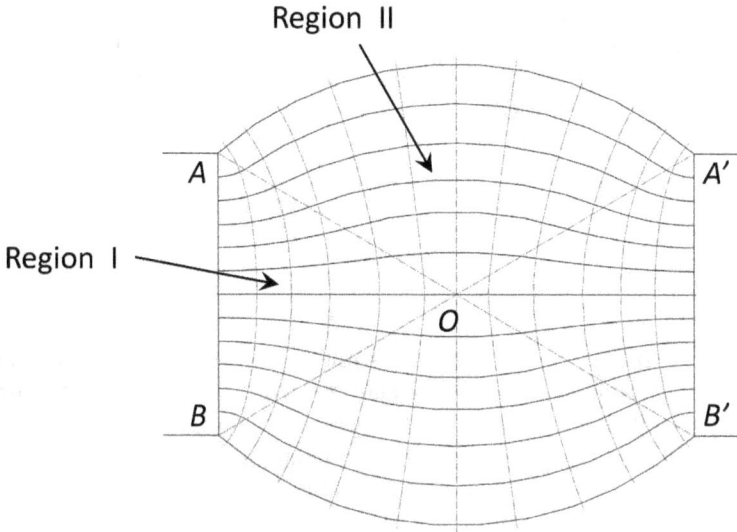

FIGURE 2.14: Coordinate systems obtained from CEC

From the point of view of the irradiance vector $\vec{D}$, we can explain fig. 2.13 in the following way. The solid lines are the field lines produced by a 2D Lambertian source with the geometry of a truncated wedge. By means of the flowline design method these field line can produce ideal devices, in this case the Compound Parabolic Concentrator using field lines which start at the vertex A or A', or HyperParabolic Concentrator using field lines which start at points betweeen A and A'. The dashed lines in the fig. 2.13 are orthogonal to the field lines and of course will play a role in the design of optical devices. If field lines define ideal reflective geometries, the geometries orthogonal to field lines can play an interesting role as refractive geometries. We will return to this interesting question in chapter 4.

In a way analogous to what we have done in the above paragraphs, it is possible to define an orthogonal coordinate system derived from the Compound Elliptical Concentrator and HyperElliptical Concentrator . It is shown in fig. 2.14. In region I, the triangle between A, B, and O, the solid lines are confocal hyperbolas and the dashed lines are confocal ellipses with focal points at A and B. In region II the solid lines are confocal ellipses and the dashed lines are confocal hyperbolas with focal points at B and B'. This type of coordinate system can be found also in conformal mapping handbooks [22] ; as we have mentioned, conformal mapping is a well-established technique to obtain 2D orthogonal coordinate systems with different geometries and boundary conditions by using the properties of analytical complex functions. The modern theory of conformal mapping is part of the theory of functions of complex variables. The foundation of the theory of analytic functions was laid by Euler [29], who was the first to obtain the conditions connecting the real and imaginary parts of the analytic complex function:

$$w = \mathfrak{F}(z), \qquad (2.8)$$

called Cauchy-Riemann conditions

$$\frac{\partial u}{\partial x} = \frac{\partial v}{\partial y}, \quad \frac{\partial u}{\partial y} = -\frac{\partial v}{\partial x}, \qquad (2.9)$$

where $w = u + iv$ and $z = x + iy$. Thus angles are preserved by the transformation $\mathfrak{F}$ and squares in the Cartesian $z-$plane always map into curvilinear squares in the $w-$plane. This conformal mapping technique has been implemented in optics [30] and also in nonimaging optics [31]. In particular the orthogonal coordinate system represented in fig. 2.14 can be obtained by mapping Cartesian coordinates by the function

$$t = \cosh z, \ w = E(t, m) \qquad (2.10)$$

where $E(t, m)$ is the elliptic integral of the second kind and $m = 0.25$; for more details on the conformal mapping technique see [22]. These examples prove the fundamental relation between coordinate systems and field theory, and of course with the field theory of nonimaging optics.

2.5   3D Hyperboloid

In previous sections of this chapter we have studied devices which are designed from the revolution of a 2D profile, like CPC or HPC. In this and the following sections we will study 3D devices working under the flowline design method, perfect mirror surfaces. 3D devices means that the mirrored surfaces of the devices follows the geometry of $\vec{D}$ produced by 3D sources. A disk as a Lambertian radiator, that is, the circular opening of a black body cavity—is the most important example to consider. The components of $\vec{D}$ can be obtained by integration of eq. 1.3 over the surface of the disk. These integrations have been carried out for radiative transfer purposes and we shall simply quote the result. Sparrow and Cess [32], for example, provide the details, and they also give a useful transformation of the surface integral over the disk into a line integral around its edge; this considerably simplifies the calculation.

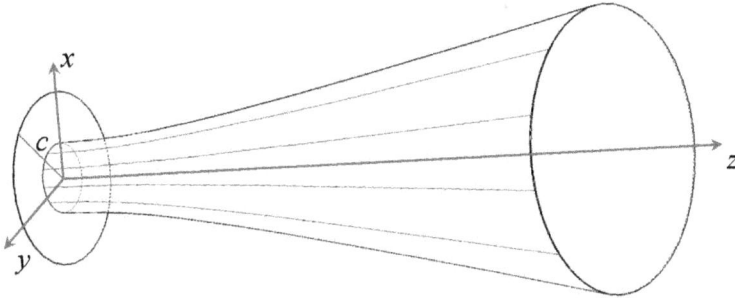

FIGURE 2.15: Geometry for calculating $\vec{D}$ of a Lambertian disk

Let c be the radius of the disk and take the coordinate axes as in fig. 2.15; then the components of $\vec{D}$ in the y, z plane, which from symmetry are all that need be considered, are given by

$$D_y = \frac{\pi}{2}\frac{z}{y}\left(\frac{c^2 + y^2 + z^2}{[(c^2 + y^2 + z^2)^2 - 4c^2y^2]^{1/2}} - 1\right) \tag{2.11}$$

$$D_z = \frac{\pi}{2}\left(\frac{c^2 - y^2 - z^2}{[(c^2 + y^2 + z^2)^2 - 4c^2y^2]^{1/2}} + 1\right). \tag{2.12}$$

The slope of the field lines is given by arctan D_y/D_z, and when this is calculated, it turns out that the field lines are the same hyperbolas as for the 2D case. The 3D surfaces are confocal hyperbolas of revolution. Mirroring the 3D hyperbolas of revolution, the behavior of the ideal device is as follows: using the reflection properties of the hyperbola, all rays incident on the entry aperture of the device (larger aperture), and directed to any point on the circular disk, will reach the exit aperture (smaller aperture) by successive reflections off the surface.

Following the coordinate system method explained in the previous section, it is easy to identify the coordinates that produce 3D hyperboloids, Oblate Spheroidal Coordinates (η, ϕ, ψ), for which the coordinates surfaces are oblate spheroids ($\eta = const$), hyperboloids of revolution ($\phi = const$) and half planes ($\psi = const$). Figure 2.16 shows the oblate spheroidal coordinates system. Note that for $\eta = 0$ the oblate spheroid becomes a circular disk, and we obtain the configuration of fig. 2.15.

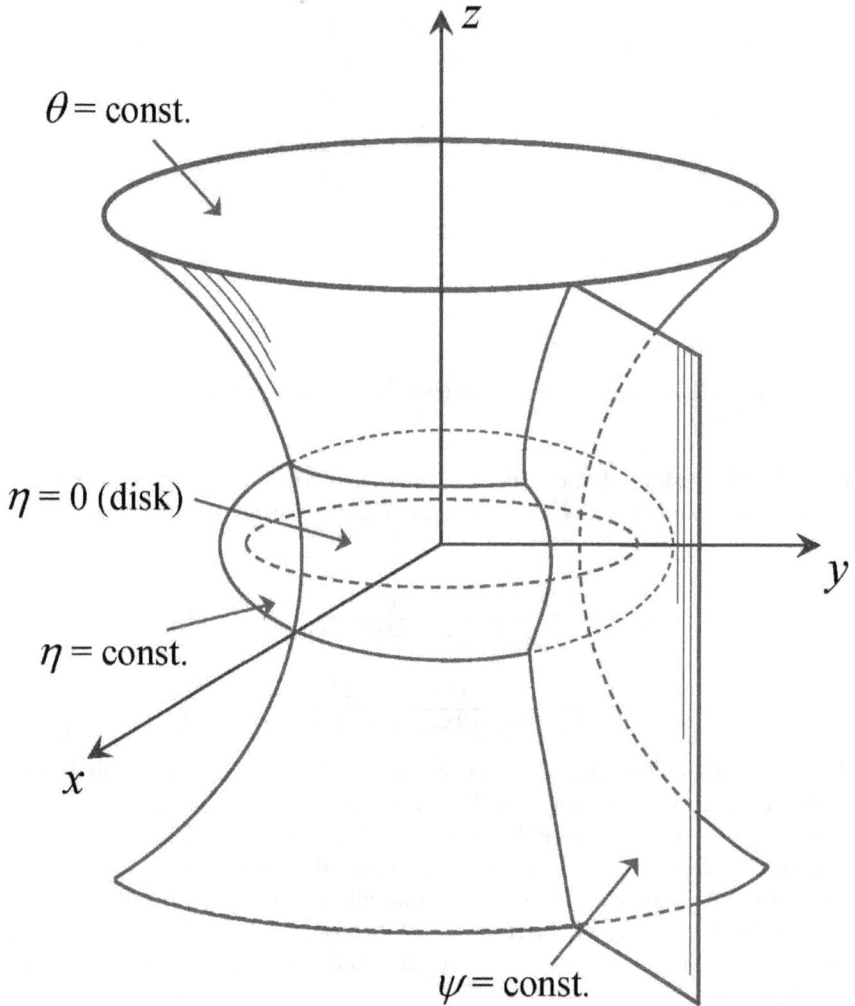

FIGURE 2.16: Oblate Spheroidal Coordinates (η, ϕ, ψ) to obtain $\vec{D}$
produced by a Lambertian disk

2.6   One-Sheeted Hyperboloid

Another interesting example of a flowline device is the one-sheeted hyperboloid, which is a 3D asymmetric ideal device [33]. In this section, we are
going to analyze its behavior from a flowline perspective, and in chapter 3

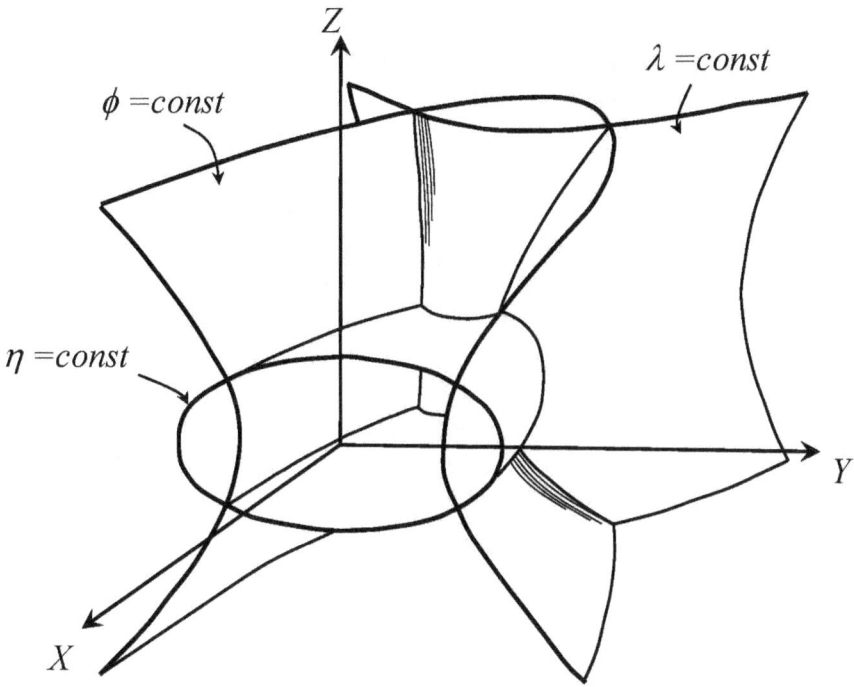

FIGURE 2.17: Ellipsoidal Coordinates (η, ϕ, λ) to obtain $\vec{D}$ produced by an elliptic Lambertian disk

we will return to the analysis of one-sheeted hyperboloids by using the vector potential of $\vec{D}$. The easiest way to study a one-sheeted hyperboloid as an optical device is by using the coordinate systems method. Let us consider the ellipsoidal coordinates of fig. 2.17, where the coordinate surfaces are ellipsoids ($\eta = const$), one-sheeted hyperboloids ($\phi = const$) and two-sheeted hyperboloids ($\lambda = const$). In way analogous to that in the previous section, for $\eta = 0$ the ellipsoid becomes an elliptic disk, and the flux tubes of $\vec{D}$ produced by a Lambertian elliptic disk have the shape of a one-sheeted hyperboloid, fig. 2.18. Again mirroring the one-sheeted hyperobolid, the behavior of the device is as follows: all rays incident on the elliptic entry aperture and directed to any point on the elliptic disk will reach the exit aperture (small elliptic section) by successive reflections, working as an ideal concentrator.

From a ray optics point of view, considering that meridional sections of a one-sheeted hyperboloid are hyperbolas and using the basic properties of a 2D hyperbolic reflector, any meridional ray incident on the elliptic entry aperture and directed to any point on the elliptic disk will reach the exit aperture. Nevertheless, to show the ideal behaviour of the one-sheeted

One-sheteed hyperboloid

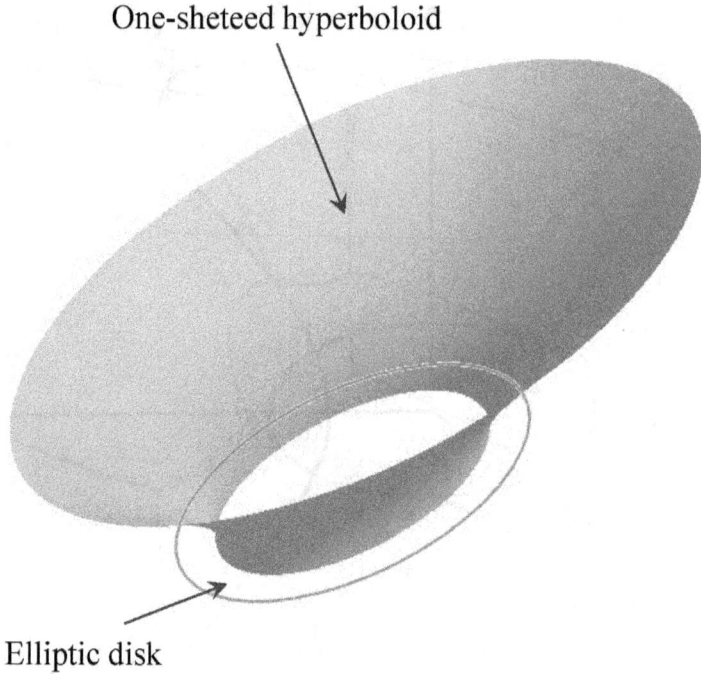

Elliptic disk

FIGURE 2.18: One-sheeted hyperboloid produced by an elliptic disk

hyperboloid, it is necessary to prove that all rays, not only meridional, directed to the elliptic disk at the exit aperture are reflected by the one-sheeted hyperboloid to some point on this elliptic disk. To do that, we can use a similar, but more general, proof than used for skew rays in hyperbolic concentrators (Winston et al,[24]). Three properties of one-sheeted hyperboloid geometry are useful. First, all meridional sections of a one-sheeted hyperboloid are hyperbolas; second, all cross sections of a one-sheeted hyperboloid are ellipses; and third, the tangent plane, at any point P of a one-sheeted hyperboloid, is defined by the bisector of the angle FPF', where F and F' are the focal points of the meridional hyperbola and the tangent to the elliptic cross section at P. All the skew rays incident at point P directed to the elliptic disk form an oblique elliptical cone, or incident cone. Then the reflected cone will be the mirror image of the incident cone, through the tangent plane at point P. By the geometry of this particular problem, the cross section of the incident cone, normal to the bisector of the angle FPF', is an ellipse. One of the principal axes of this ellipse lies in the tangent plane of the one-sheeted hyperboloid and, by definition, the bisector of the incident cone too; therefore,

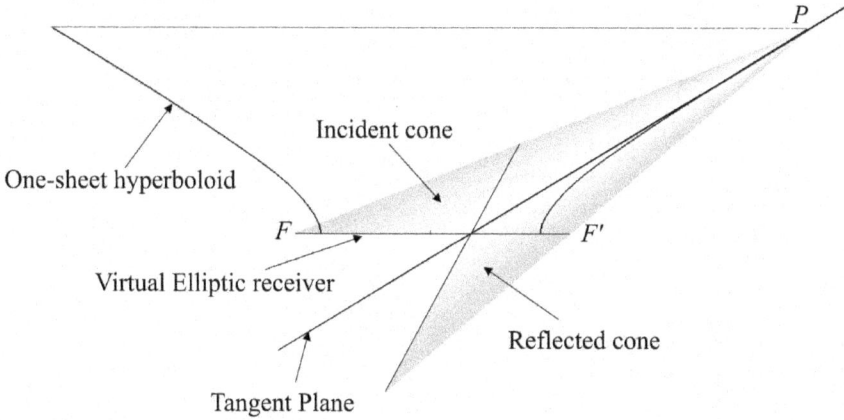

FIGURE 2.19: Incident cone, reflected cone and tangent plane at
one-sheeted hyperboloid

the tangent plane coincides with a symmetrical plane of the incident cone,
fig. 2.19. The result is that the reflected cone coincides with the incident cone,
which means that all rays incident at point P directed to the elliptic disk are
reflected by the one-sheeted hyperboloid to some other point on the elliptic
disk. This proves that a one-sheeted hyperboloid is an ideal 3D asymmetric
device.

To show the behaviour of the one-sheeted hyperboloid we have analyzed its
efficiency with raytracing simulations. To do that, it is necessary to quantify
the number of rays linking the input and output surfaces with a perfect mirrored device and to compare it to the number of rays linking the input surface
and the elliptic disk at each incident angle, so we need to perform two raytrac-
ings by each measure (angle). To perform these simulations we have employed
a perfect one-sheeted hyperboloid mirror, reflectance $\rho = 1$, considering the
equation of the one-sheeted hyperboloid in Cartesian coordinates,

$$\frac{x^2}{a^2} + \frac{y^2}{b^2} - \frac{z^2}{c^2} = 1, \tag{2.13}$$

with parameters $a = 50\,mm$, $b = 25\,mm$, $c = 30\,mm$ and height of
$z_{max} = 70\,mm$. We have traced a beam of collimated rays, which fills the
entry aperture of the one-sheeted hyperboloid, for different incident angles.
For this beam we have measured on the one hand, the number of rays incident to the elliptic disk without the one-sheeted hyperboloid, and on the other
hand, the number of rays emerging from the exit aperture of the one-sheeted
hyperbolic concentrator after reflection. Figure 2.20 shows the ratio between

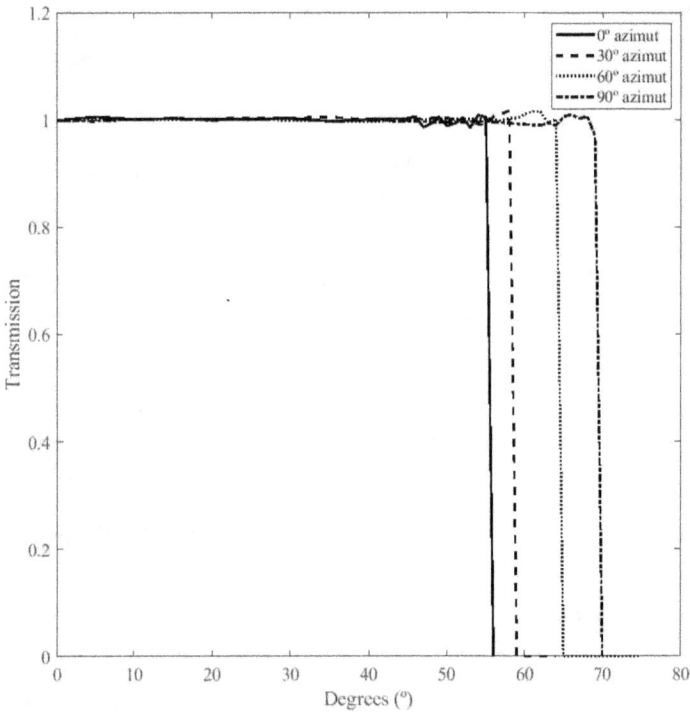

FIGURE 2.20: Transmission-angle curves for a one-sheeted hyperbolic
concentrator, reflectance 1, at several azimuth incidence angles

these measurements for all incident angles and for different azimuth angles of incidence. Using the parameters of the employed concentrator, the cutoff angle at the minor semiaxis incident is $\theta_{max} = 55.7°$ and at the major semi-axis is $\theta_{max} = 69.3°$. Figure 2.20 shows also the ideal 3D rectangular cutoff behaviour of the one-sheeted hyperbolic concentrator, for azimuth angles of incidence being $0°$; the graph shows the result of these rays according to their angle in the plane of the minor semiaxis, showing the concentrator's ideal 3D asymmetrical behaviour. Obviously for real reflectance values, around 0.95, the transmission angle curve will reduce its efficiency. It shows too the small simulation errors, which appear near the cutoff angle due to the faceted design of the concentrator. For nontracking solar concentration applications, infinite source concentrators are needed; then to obtain an infinite source concentrator an asymmetric lens must be added at the entry aperture of the one-sheeted hyperbolic concentrator, in a similar way to that done by O'Gallagher et al. [34]. To conclude, a one-sheeted hyperboloid is an ideal 3D asymmetric concentrator as its shape do not disturb the field lines of an elliptic disk. In general it can be applied for nontracking solar concentration where two different acceptance angles, at transversal and longitudinal directions, are needed.

2.7 Flowline Asymmetric Nonimaging Concentrating Optics

Using flowline and its property of guiding étendue within the ideal concentrators, we can have another degree of freedom in the design when it comes to certain restrictions of concentrator (or illuminator) designs. In a standard setup of a sun concentration system, not only the position of the sun is predetermined relative to the position of the absorber, e.g. due to the local latitude; the tilting of the aperture of the concentrator can also be limited due to restrictions, such as shading, or the covering glass [35]. As an example, we can consider that the building-integrated PV module (BiPV) may also require the concentrator aperture to be parallel to the wall, in order to minimize the shading between concentrators. By searching among the flowlines within the ideal concentrator BC, $B'C'$ (fig. 2.21), we can meet such a requirement by limiting the aperture to be parallel to the absorber. A simple binary search routine using starting points C_o, C_1,... for flowlines is shown in.

In this process, the tilting of aperture $B'B_o$, $B'B_1$... etc, is compared with the angle of CC', and the program stops when the angle difference is within the tolerance of the design. This results in the concentrator shown in fig. 2.22.

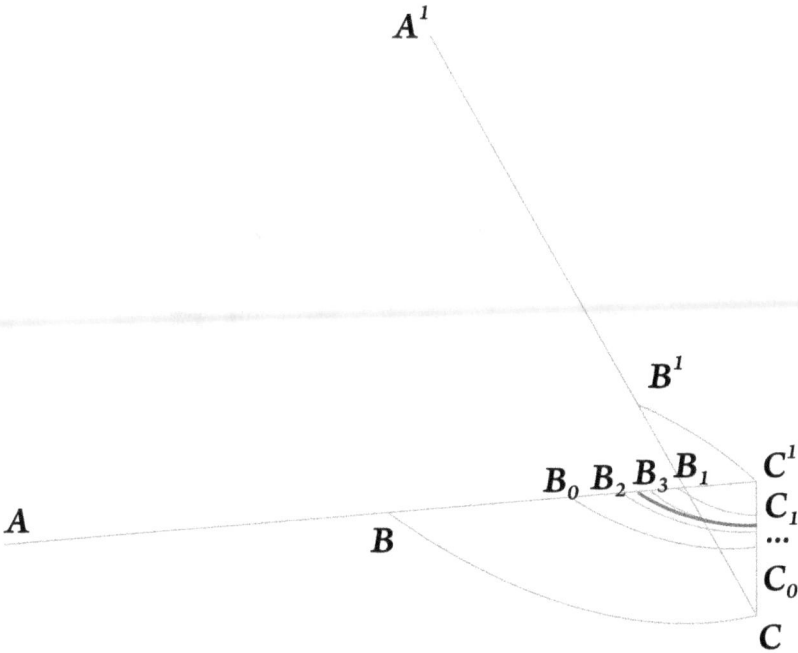

FIGURE 2.21: By adjusting the starting position of the flowline within the absorber CC', we can adjust the angle of aperture B'

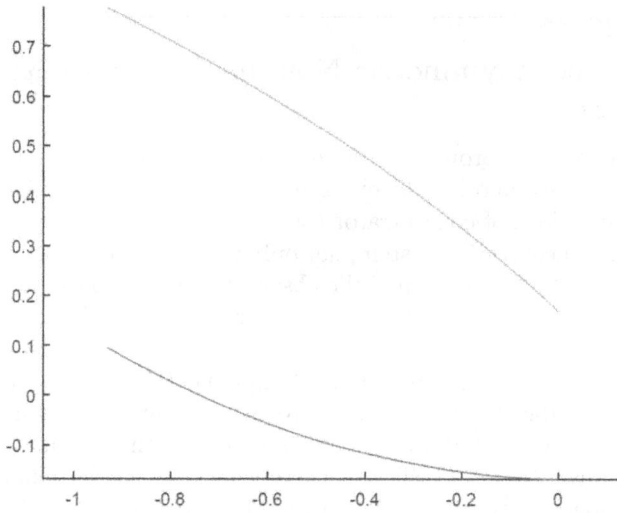

a) Asymmetric concentrator based on flowline

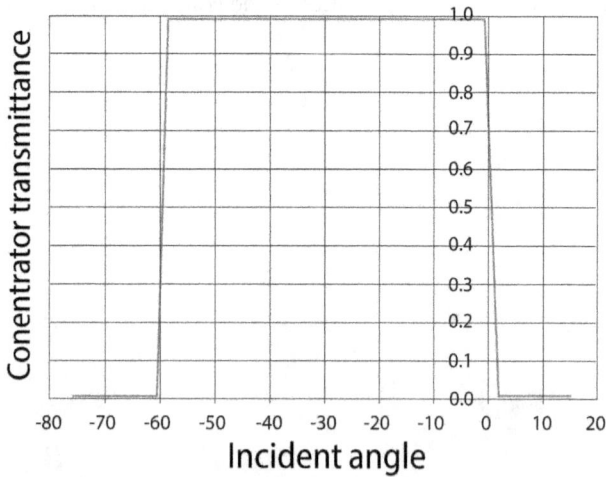

b) Asymmetric transmission curve

FIGURE 2.22: The optical simulation of an ideal, nonimaging, asymmetric, flowline design, which meets the requirement of the aperture being parallel to the absorber

By constructing an array of such concentrators, not only the relevant étendue at the aperture (the seasonal angle variation of the sun in this case, according to the full area of the wall) is fully used, but the ideal concentration law of

$$C_{max} = \frac{1}{F_{12}} \tag{2.14}$$

where F_{12} is the view factor, is also satisfied. The flowline in this case provides another degree of freedom to the ideal concentrator design by allowing the tilting angle of the aperture to also be flexible. Such a result cannot be achieved by simply tilting the conventional CEC. In fig. 2.23 we showcase the flowline of absorbers with an arbitrary convex shape. The potential of flowline is demonstrated here as being capable of providing another degree of freedom for any existing CPC/CEC designs.

2.8   Ideal Source-Receiver Transmission Design

Considering that the flowline method provides a technique to perfectly transport the energy from the entry aperture of a device to the exit aperture, an interesting question arises: is it possible to theoretically design a device which perfectly transport all the radiation produced by a Lambertian disk source to a disk , of the same size, in any other place in space? A perfectly mirrored cylinder provides the trivial answer for an aligned source and receiver. The answer for the general problem of perfectly connecting a disk source with a disk receiver of the same size placed in any location in space is yes, and the way to demonstrate this comes again from coordinate systems.

As we have mentioned, any coordinate system can be viewed as a flowline device. In this particular case what we need to do is to freely move a disk in space following coordinate systems (surfaces). We can do that considering toroidal coordinates, fig. 2.24, and cylindrical coordinates toroidal coordinates will produce curvilinear movement of the disk and cylindrical coordinates will produce rectilinear movement of the disk. The situation is similar to the movement of a particle: the toroidal coordinates produce circular movement of the disk, but any curvilinear movement can be obtained by successive infinitesimal circular movements, which means succesive toroidal coordinates.

In fig. 2.25 we show an example of a device to perfectly connect a disk source to a disk receiver. The first step is to design a toroid with the source disk. In fig. 2.25a we show a toroid with angle π divided in two parts; the exit surface of this first part of the device is the source for the second part of the

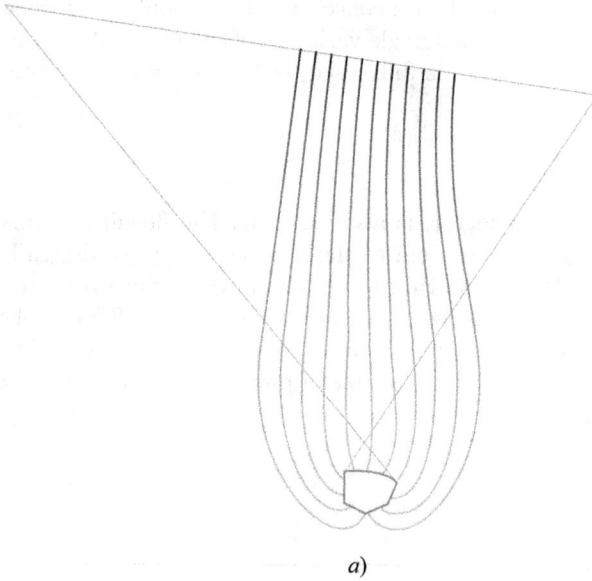

a)

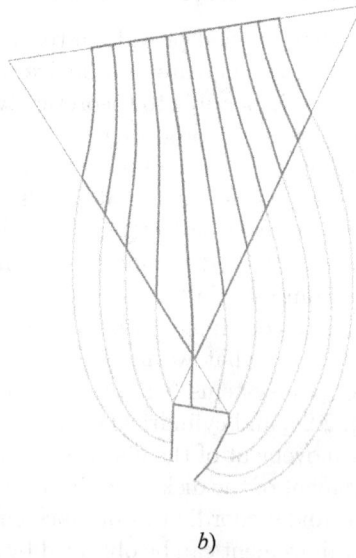

b)

FIGURE 2.23: An asymmetric flowline design can be generalized into any convex shape, a) closed shape, and b) open shape

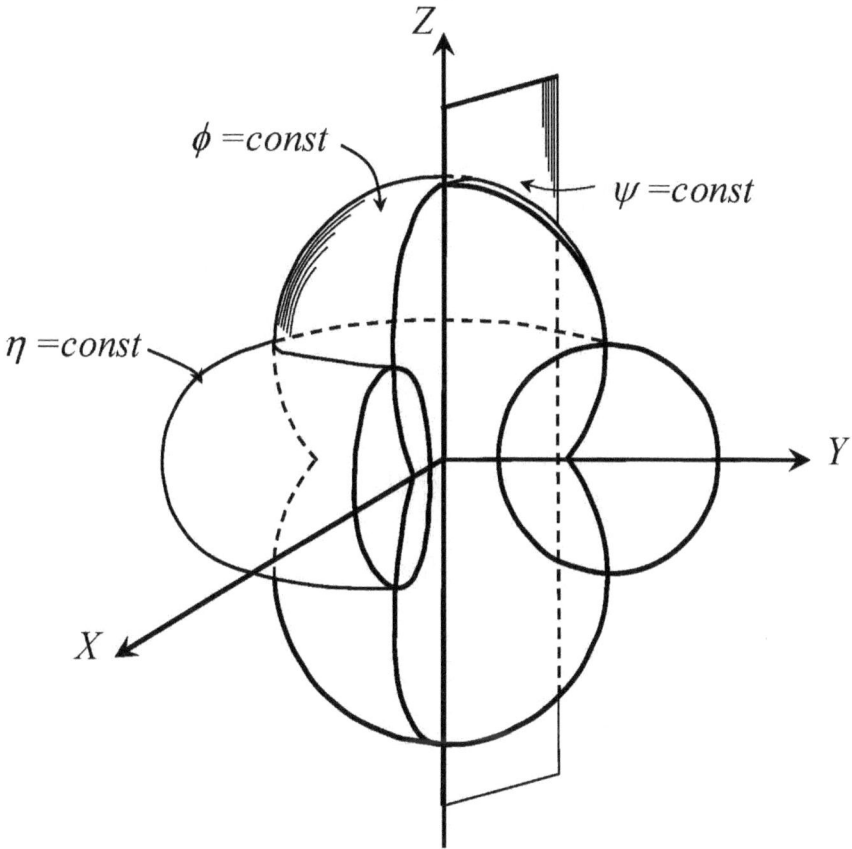

FIGURE 2.24: Toroidal coordinates (η, ϕ, ψ), for which the coordinate surfaces are toroids $(\eta = const)$, spherical bowls $(\phi = const)$ and half planes $(\psi = const)$

device. Figure 2.25b shows the device, with the second component rotated; the rotation axis passes through the center of the exit surface of the first component and is normal to the exit surface; in this case the angle of the tilt is 45°. Following this procedure it is possible to consider the exit surface of the second component as the source for the third component, and then we can rotate or translate the Lambertian disk freely in space.

Rotation axis

a)

Rotation axis

b)

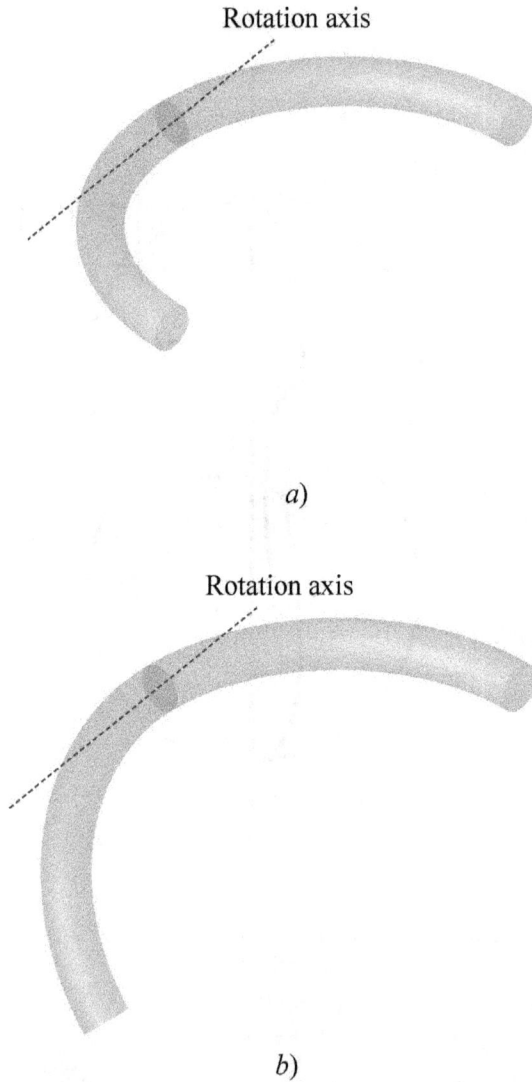

FIGURE 2.25: a) A toroidal ideal device, b) with successive rotated
toroidal elements

Finally in fig. 2.26 we show a raytracing for a successive perfectly mirrored
toroidal device producing a complete loop; the raytracing checks that all rays
from the Lambertian source reach the exit aperture after successive reflections
in the device, with no rays going back.

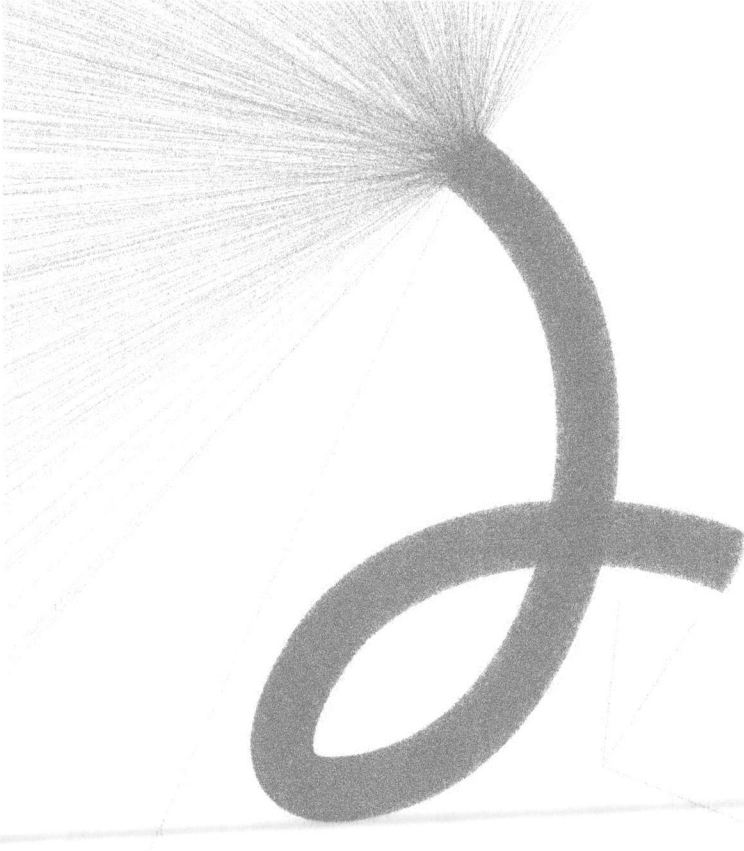

FIGURE 2.26: Raytracing of a toroidal ideal device with a complete loop, no turn back rays

3

Field Theory Elements

The task of this chapter is to discuss the mathematical techniques which are useful in the calculation and analysis of free space propagation of $\vec{D}$. An interesting application of contour integrals will be studied in section 3.4; it provides an easy and fast way to compute the $\vec{D}$ for planar sources with arbitrary contours. We also introduce and apply the concepts of vector potential, quasipotential and gauge invariance of $\vec{D}$. In the last two sections, using the fundamental concept of the cone edge rays introduced in section 1.5, we apply Lorentz geometry to the computation of $\vec{D}$. Gutierrez et al. [36], using the flowline design method, applied Lorentz geometry to obtain rotational symmetric concentrators. In this chapter we are going to apply Lorentz geometry to compute the direction and the modulus of $\vec{D}$ produced by symmetric and non symmetric radiation sources.

3.1 Classification of Nonimaging Optics Fields

In Chapter 1 we defined the irradiance vector $\vec{D}$ and introduced some basic properties of the vector field produced in free space, and we studied the divergence and curl of $\vec{D}$. Now we are going to study some more fundamental properties of the vector field, such as the vector potential, gauge invariance and quasipotential, and we will obtain interesting properties of orthogonal coordinate systems. We can start our analysis with a basic classification of vector fields in terms of divergence and curl. From a mathematical point of view it is interesting to consider the Helmholtz decomposition theorem [37], which states that any vector field $\vec{D}$ can be expressed as the sum of two parts, one of which is irrotational and the other of which is solenoidal

$$\vec{D} = -\nabla\phi + \nabla \times \vec{A}, \tag{3.1}$$

then

$$\nabla \cdot \vec{D} = -\nabla \cdot \nabla\phi + \nabla \cdot \nabla \times \vec{A} = -\nabla^2\phi, \tag{3.2}$$

DOI: 10.1201/9780367551605-3

which provides a differential equation to compute scalar potential ϕ, and

$$\nabla \times \vec{D} = -\nabla \times \nabla \phi + \nabla \times \nabla \times \vec{A} = \nabla \times \nabla \times \vec{A} = \nabla(\nabla \cdot \vec{A}) - \nabla^2 \vec{A}, \quad (3.3)$$

which provides a second partial differential equation to compute vector potential $\vec{A}$, then appears to naturally classify the field of $\vec{D}$ in terms of scalar and vector potential. Also we must take into account that in free space, without absorption or emission of radiation, $\nabla \cdot \vec{D} = 0$.

The first class of fields is *Conservative Fields*, the necessary and sufficient condition for the existence of a scalar potential ϕ defined by

$$\vec{D} = -\nabla \phi \quad \text{is} \quad \nabla \times \vec{D} = 0. \tag{3.4}$$

This situation only occurs in a few special vector fields produced by $\vec{D}$: the field produced by a point source, the field produced by a perfect spherical source or a perfect infinite plane source; those fields are called irrotational fields.

The second class is *Quasipotential Fields*, the necessary and sufficient condition for the existence of a quasipotential Φ_q defined by

$$\vec{D} = -\frac{1}{\mu} \nabla \Phi_q \quad \text{is} \quad \vec{D} \cdot \nabla \times \vec{D} = 0 \tag{3.5}$$

where μ is an scalar integrating function which depends on the position. This is the case for most practical situations, and we will study it in detail in section 3.7 of this chapter.

The third class is *Solenoidal Fields*, the necessary and sufficient condition for the existence of a vector potential $\vec{A}$ defined by

$$\vec{D} = \nabla \times \vec{A} \quad \text{is} \quad \nabla \cdot \vec{D} = 0. \tag{3.6}$$

This is valid for all geometries of light sources, which gives the vector potential a special relevance in the study of the vector field $\vec{D}$. The study of the vector potential of $\vec{D}$ is inherently harder than other techniques, but it is a robust technique in the sense that it can be applied to most cases, and theoretically it can produce a partial differential equation, the vector Poisson equation, which at least can produce a numerical solution for any particular situation.

3.2   Geometrical Description of Modulus of $\vec{D}$

The flowline method uses the direction of $\vec{D}$ to design optical devices but does not provide information about the modulus of $\vec{D}$. In this section, we study some concepts and techniques to evaluate the modulus of $\vec{D}$ for Lambertian sources . Nonimaging optics typically considers two basic symmetries, 2D or translational symmetry and 3D or rotational symmetry devices [24]. From this point of view, a 2D Lambertian source is an infinite-length rectangular source, and a 3D Lambertian source is a disk source. It is a well-known result that for 2D sources [4][24], the $\vec{D}$ at any point P in space points along the direction of the bisector of the edge rays and has the modulus

$$|\vec{D}| = B \sin \theta \tag{3.7}$$

where θ is the bisector angle of the edge rays and B is the brightness. It is possible to generalize this result to 3D sources. The modulus of a 3D Lambertian source is also a well-known result [5], and it is interesting that at any point P on the axis of the disk the modulus of $\vec{D}$ is

$$D = B \sin^2(\theta) \tag{3.8}$$

where θ, in this case, is the semiangle of the cone of edge rays produced at point P on the axis.

These two results appear to have no connection between them, but it is possible to establish a connection using symmetry arguments. From a coordinate point of view, a 2D Lambertian source can be easily represented using Cartesian coordinates, while a 3D Lambertian source can be easily represented using polar coordinates. Now take a look at elliptical coordinates in fig. 3.1, where c is the focal length of the confocal hyperbolas and ellipses. In the limit

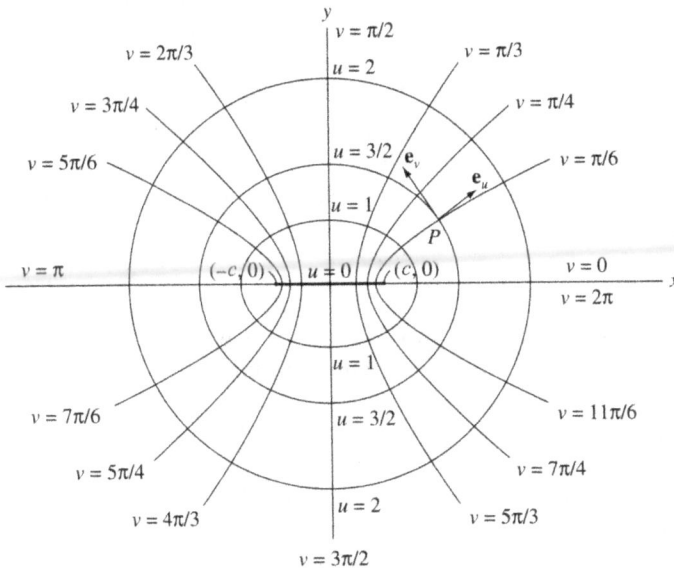

FIGURE 3.1: Elliptical coordinates (u, v); for $c \to 0$ they transform to polar coordinates and for $c \to \infty$ they transform to Cartesian coordiantes

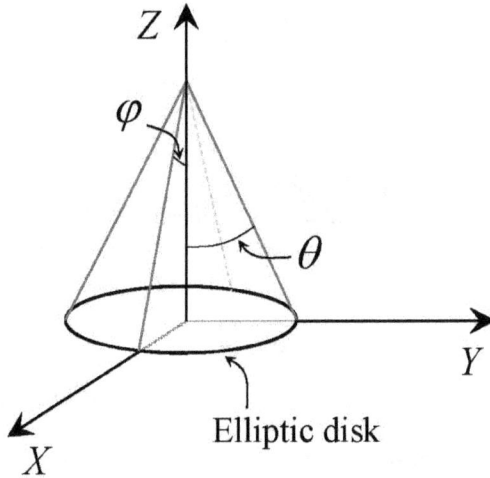

FIGURE 3.2: Angle configuration for the calculation of modulus of $\vec{D}$ for an elliptic Lambertian disk source

$c \to 0$ the elliptical coordinates in fig. 3.1 transform to polar coordinates. On the other hand, in the limit $c \to \infty$ the elliptical coordinates transform to Cartesian coordinates [38]; in this way we can see polar coordinates and Cartesian coordiantes as limit cases of elliptical coordinates. In an analogous way 2D Lambertian sources and 3D Lambertian sources can be considered as limit cases of elliptic Lambertian disk sources. As a consequence of this symmetry, eq. 3.7 and eq. 3.8 can be considered as limit cases for the modulus of $\vec{D}$, at any point P in the axis of an elliptic Lambertian disk source, which can be written in the form

$$D = B \sin(\theta) \sin(\varphi) \tag{3.9}$$

where φ is the angle of the elliptical cone of edge rays through the direction of focal length c, fig. 3.2. Note that symmetry limits are fulfilled by eq. 3.9: in the limit $c \to \infty$, $\sin(\varphi) \to 1$, eq. 3.9 transforms to eq. 3.7; on the other hand, in the limit $c \to 0$, $\sin(\varphi) \to \sin(\theta)$, eq. 3.9 transforms to eq. 3.8.

We have checked eq. 3.9 by raytracing simulations ; fig. 3.3 shows the comparison between raytracing and the modulus of $\vec{D}$ obtained by eq. 3.9 introduced in this section, at the axis of Lambertian sources, for a circular disk and for an elliptic disk, showing agreement.

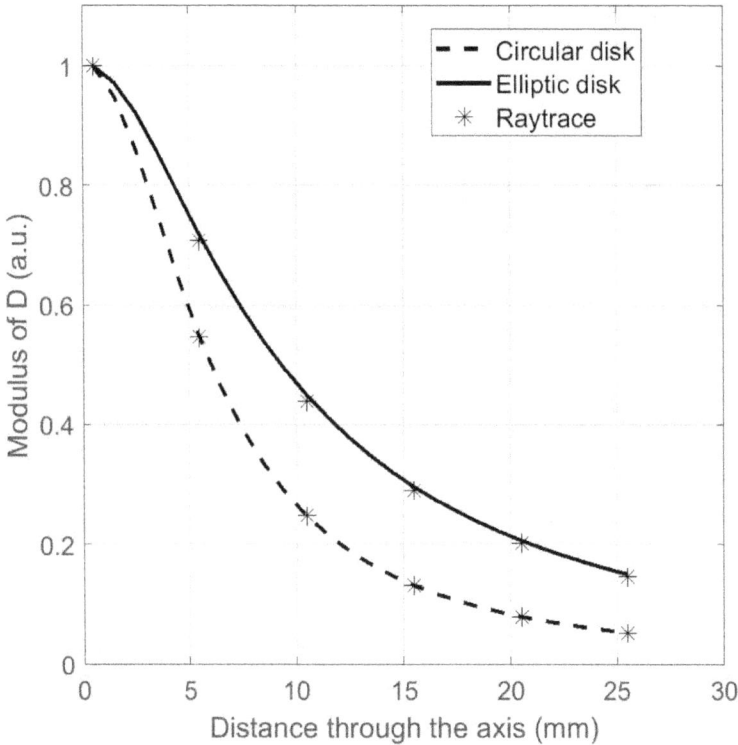

FIGURE 3.3: Modulus of $\vec{D}$ at the axis of a circular Lambertian disk and an elliptic Lambertian disk

The procedure presented in this section can be generalized to points outside the Z axis. In fact, for a circular source at any point in space, P, it is possible to define an elliptic light cone, the cone of edge rays, but we will need to define a new local basis at point P $(\vec{X}',\vec{Y}',\vec{Z}')$ with one axis in the direction of the axis of the cone, and the other two orthogonal axes parallel to the axis of the elliptical base of the cone of edge rays, fig. 3.4. We will return to this concept using Lorentz geometry formalism at the end of this chapter.

3.3 Contour Integrals of the Light Field

An interesting method of calculating the $\vec{D}$ field appears when we replace the traditional surface integral eq. 1.3 of a source of constant brightness with a

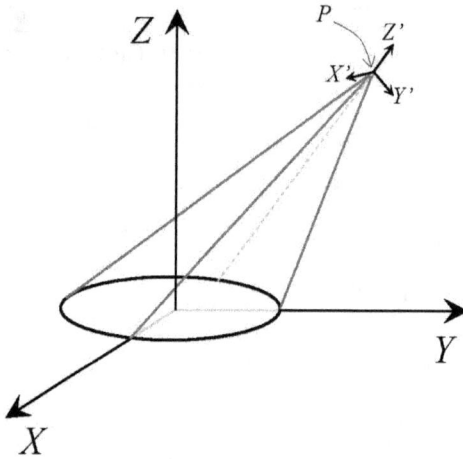

FIGURE 3.4: Sketch to compute $\vec{D}$ outside the axis of a circular
Lambertian disk

contour integral around the boundary of the source. We can do that using
Stoke's theorem, which relates the surface integral to the contour integral of
any arbitrary vector $\vec{D}$:

$$\int_s \nabla \times \vec{D} \cdot d\vec{S} = \oint_l \vec{D} \cdot d\vec{l}. \tag{3.10}$$

Stoke's theorem was first applied to the computation of the irradiance vector by Fock [3],[39], who used contour integrals to evaluate radiation transfer
between a source and an absorber. Moon [5] developed it to compute $\vec{D}$ produced by polygonal sources. Consider a source S of uniform brightness B: the
surface need not be plane, but it must be such that its complete boundary
can be seen from every point P in space, for which we want to compute $\vec{D}$

$$\vec{D} = \int_s \frac{B\vec{u}_r}{r^2} (\vec{u}_r \cdot d\vec{S}) = \int_s \frac{B\cos(\theta)\vec{u}_r}{r^2} dS \tag{3.11}$$

where $\vec{u}_r$ is a unit vector in the direction of $\vec{r}$. Multiplication by an arbitrary
unit vector $\vec{N}$ gives

$$\vec{N} \cdot \vec{D} = \int_s \frac{B}{r^2} (\vec{N} \cdot \vec{u}_r)(\vec{u}_r \cdot d\vec{S}) \tag{3.12}$$

$$\vec{N} \cdot \vec{D} = B \int_s \left(\frac{\vec{u}_r}{r^2} (\vec{N} \cdot \vec{u}_r) \right) \cdot d\vec{S}. \tag{3.13}$$

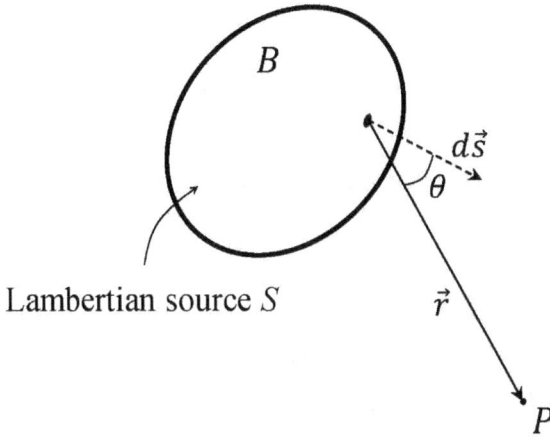

FIGURE 3.5: Computation of $\vec{D}$ from source S

To introduce Stoke's theorem, we must express the integrand of eq. 3.13 as the *curl* of something. Gershun [4] found that

$$\frac{1}{2}\nabla \times \left(\frac{\vec{u}_r}{r} \times \vec{N}\right) = \frac{\vec{u}_r}{r^2}(\vec{N} \cdot \vec{u}_r); \tag{3.14}$$

therefore

$$\vec{N} \cdot \vec{D} = \frac{B}{2}\int_s \nabla \times \left(\frac{\vec{u}_r}{r} \times \vec{N}\right) \cdot d\vec{S}, \tag{3.15}$$

or by Stoke's theorem,

$$\int_s \nabla \times \left(\frac{\vec{u}_r}{r} \times \vec{N}\right) \cdot d\vec{S} = \oint_l \left(\frac{\vec{u}_r}{r} \times \vec{N}\right) \cdot d\vec{l}; \tag{3.16}$$

thus,

$$\vec{N} \cdot \vec{D} = \frac{B}{2}\oint_l \left(\frac{\vec{u}_r}{r} \times \vec{N}\right) \cdot d\vec{l}, \tag{3.17}$$

and since the terms of a scalar triple product may be commuted cyclically,

$$\vec{N} \cdot \vec{D} = \vec{N} \cdot \frac{B}{2}\oint_l d\vec{l} \times \frac{\vec{u}_r}{r} \tag{3.18}$$

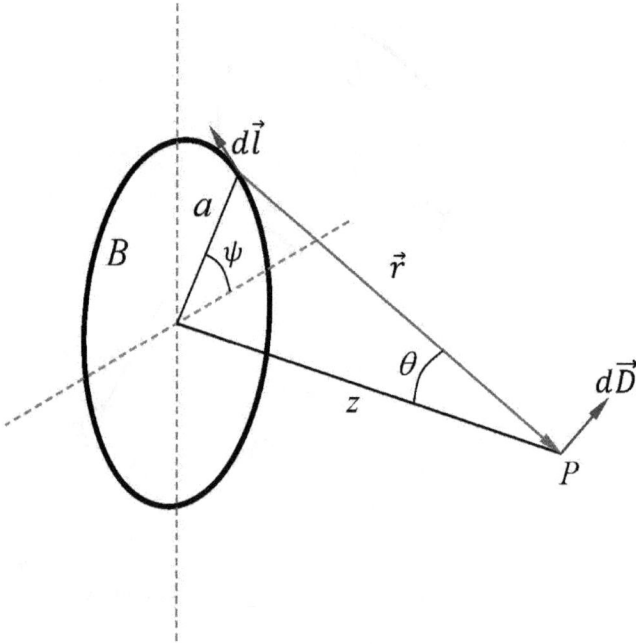

FIGURE 3.6: Contour evaluation of $\vec{D}$ at a point P on the axis of a circular Lambertian source, by applying eq. 3.19

or

$$\vec{D} = \frac{B}{2} \oint_l d\vec{l} \times \frac{\vec{u}_r}{r}.$$ (3.19)

This equation gives a very interesting way to obtain $\vec{D}$ from a surface of constant brightness.

As a simple example of contour computation of $\vec{D}$, consider a circular disk of radius a and brightness B, and a point P on the axis, fig. 3.6. For each element $d\vec{l}$ we have an element $d\vec{D}$; the radial components of $d\vec{D}$ cancel in integrating around the contour, so we need to consider only the z component. Then

$$\vec{D} = \frac{B}{2} \vec{k} \oint \frac{dl}{r} \sin \theta$$ (3.20)

and also

$$\frac{dl}{r} = a\frac{d\psi}{r}, \tag{3.21}$$

where ψ is the angle about the z axis; therefore

$$\vec{D} = \frac{aB\sin(\theta)}{2r}\vec{k}\oint_0^{2\pi} d\psi. \tag{3.22}$$

Considering that

$$\frac{a}{r} = \sin(\theta), \tag{3.23}$$

we have

$$\vec{D} = B\pi\sin^2(\theta)\vec{k} \tag{3.24}$$

in agreement with previous results.

Another important contour integral is related to the flux of the irradiance vector through a closed surface

$$\Phi = \oint_s \vec{D}\cdot d\vec{S}, \tag{3.25}$$

which is equal to the difference between the quantity of radiation entering the volume enclosed by the surface S per unit time, and the quantity of radiation leaving the volume. The $d\vec{s}$ has the direction of the outer normal to the surface. Thus Φ represents the radiation produced (or absorbed) in unit time in the volume v. Using Gauss' theorem

$$\oint_s \vec{D}\cdot d\vec{s} = \int_v \nabla\cdot\vec{D}dv, \tag{3.26}$$

which produces the important consequence that for all points in space for which there is no emission or absorption of radiation

$$\nabla\cdot\vec{D} = 0, \tag{3.27}$$

the irradiance vector is solenoidal and a vector potential $\vec{A}$ exists.

3.4 $\vec{D}$ Produced by an Arbitrary Plane Source

From eq. 3.19 it is possible to develop a simple contour calculation method of $\vec{D}$, which can be generalized for sources of arbitrary shape. The magnitude of the integrand of eq. 3.19 is

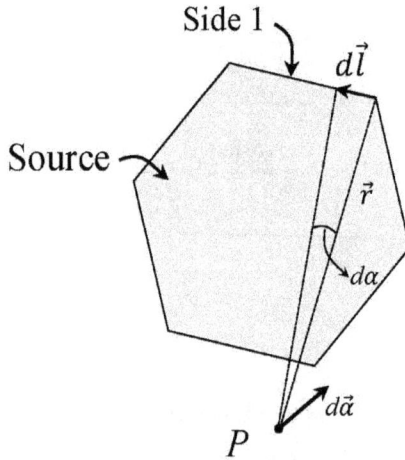

FIGURE 3.7: Contour computation of $\vec{D}$ at P produced by a
polygonal source

$$\left| \vec{dl} \times \frac{\vec{u}_r}{r} \right| = \frac{dl \sin \theta}{r} = d\alpha, \tag{3.28}$$

which is the value of the infinitesimal angle subtended by dl at P, and its
direction is orthogonal to the triangle with sides dl and r, fig. 3.7. Thus, it is
convenient to introduce the angle vector $d\vec{\alpha}$ at point P

$$\vec{dl} \times \frac{\vec{u}_r}{r} = d\vec{\alpha}, \tag{3.29}$$

and the eq. 3.19 becomes

$$\vec{D} = \frac{B}{2} \oint_l d\vec{\alpha}. \tag{3.30}$$

Now consider a polygonal source, fig. 3.7: all the vectors $d\vec{\alpha}$ are in the same
direction for each side of the polygon and add directly to form a single vector
α. As an example, for side 1 of fig. 3.7, the vector α is orthogonal to the
triangle determined by that side and point P; thus the contribution to $\vec{D}$ of
that side is

$$\vec{D}_1 = \frac{B}{2} \int_0^{\alpha_1} d\vec{\alpha} = \frac{B\alpha_1}{2} \vec{n}_1 \tag{3.31}$$

where $\vec{n}_1$ is a unit vector that is orthogonal to the triangle determined by side 1 and point P. Using similar calculations for the other sides of the polygonal source, we can obtain the $\vec{D}$ at point P as the sum

$$\vec{D} = \frac{B}{2} \sum_{i=1}^{N} \alpha_i \vec{n}_i. \tag{3.32}$$

Equation 3.32 provides a simple method to compute $\vec{D}$, not only for polygonal sources. Generalization to circular or other plane sources can be done by approaching the contour of the source with a polygonal source; increasing the number of sides of the polygon will increase the resolution of the computation. Equation 3.32 also simplifies computations: for square sources $\vec{D}$ is obtained by computing only four vectors and four angles, which can be a great advantage in optical design and simulation techniques.

As an example of the application of eq. 3.32 we can compute the irradiance produced by a circular Lambertian source of radius $10\,mm$ at a plane parallel to the circular source at a distance of $d = 0.5\,mm$ for different numbers of sides of the circular contour. In fig. 3.8 we show the irradiance for different numbers of sides of the circular contour: a) 12 sides, b) 18 sides, c) 24 sides, and d) 30 sides. This shows how it is possible to increase the accuracy of the irradiance distribution by increasing the number of sides of the contour.

3.5 Vector Potential and Gauge Invariance

Let us introduce the concept of vector potential . If the irradiance vector $\vec{D}$ is solenoidal, $\nabla \cdot \vec{D} = 0$, the flux of $\vec{D}$ through a surface S may be expressed according to Stoke's theorem, as a line integral of another vector $\vec{A}$ over a contour C embracing this surface:

$$\int_s \vec{D} \cdot \vec{ds} = \oint_l \vec{A} \cdot \vec{dl}. \tag{3.33}$$

The vector $\vec{D}$ is the *curl* of the vector $\vec{A}$, which is called the vector potential:

$$\vec{D} = \nabla \times \vec{A}. \tag{3.34}$$

The value of the concept of the vector potential resides in the fact that it allows a simplified computation of flux from one surface to another. The integral of the irradiance vector over a surface is reduced to the integral of the

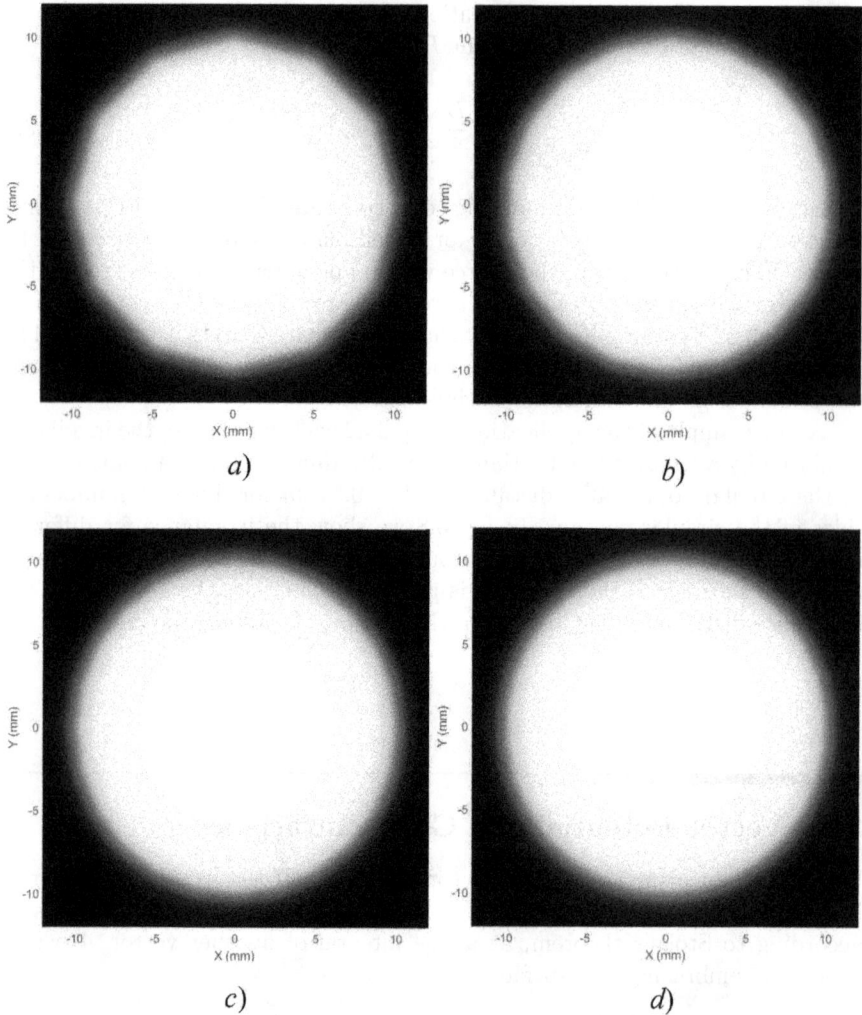

FIGURE 3.8: Irradiance computation using eq. 3.32 for a circular source with a) 12 sides in the contour, b) 18 sides in the contour, c) 24 sides in the contour and d) 30 sides in the contour

vector potential along a contour. Let's remember the well-known relationships between the components of the $\vec{D}$ and $\vec{A}$ in Cartesian coordinate systems:

$$D_x = \frac{\partial A_z}{\partial y} - \frac{\partial A_y}{\partial z}, \tag{3.35}$$

$$D_y = \frac{\partial A_x}{\partial z} - \frac{\partial A_z}{\partial x}, \tag{3.36}$$

$$D_z = \frac{\partial A_y}{\partial x} - \frac{\partial A_x}{\partial y}. \tag{3.37}$$

These relations determine the components D_x, D_y and D_z when the values of A_x, A_y and A_z are known at all points of the field. We consider now the inverse problem of finding the vector potential for a uniformly luminous surface which radiates according to Lambert's law, taking into account that

$$\frac{\vec{u}_r}{r} = \nabla \ln r. \tag{3.38}$$

Equation 3.19 may be written

$$\vec{D} = -\frac{B}{2} \oint_l \nabla \ln r \times \vec{dl} \tag{3.39}$$

using again the property of eq. 1.28

$$\nabla \times (\psi \vec{A}) = \psi (\nabla \times \vec{A}) + (\nabla \psi \times \vec{A}) \tag{3.40}$$

Equation 3.39 becomes

$$\vec{D} = -\frac{B}{2} \oint_l \nabla \times (\ln r \vec{dl}) = \nabla \times \left(-\frac{B}{2} \oint_l \ln r \vec{dl} \right), \tag{3.41}$$

where we can obtain the following expression for the vector potential

$$\vec{A} = -\frac{B}{2} \oint_l \ln r \vec{dl}. \tag{3.42}$$

We will apply this equation in the following sections and chapters to several examples.

Now we would like to consider another interesting field theory element, what is known as gauge invariance [40]. The term gauge invariance refers to the property that a whole class of scalar and vector potentials, related by so-called gauge transformations , can describe the same nonimaging optics field. It is due to the fact that scalar and vector potential are related to $\vec{D}$ by a differential equation. Let us consider the vector potential $\vec{A}$, related to $\vec{D}$ by equation eq. 3.34, considering that for any scalar field Ω

$$\nabla \times \nabla \Omega = 0, \tag{3.43}$$

it is possible to define a new vector potential $\vec{A}'$

$$\vec{A}' = \vec{A} - \nabla \Omega, \tag{3.44}$$

which provides the same vector field $\vec{D}$ at any point in space. Equation 3.44 is the so-called gauge transformation, and in some particular cases the gauge term plays an important role. A similar situation appears for scalar potential ϕ: it is possible to add a term, for example a constant, which does not modifies the vector $\vec{D}$. This nonuniqueness of the potentials gives us the possibility of choosing them so that they fulfill one auxiliary condition chosen by us. Gauge invariance can also be applied to optical path length and Fermat's principle; for example, when adding a constant to the optical path length, Fermat's principle remains unchanged . It has also been applied to relativistic and quantum fields [41].

Since vector field $\vec{D}$ can be used to produce ideal nonimaging optics, by means of flowline method, as shown in section 2.7, it is important to investigate what its vector potential is like in ideal nonimaging optics. Note that due to the nature of vector potential being gauge invariant, there are infinite numbers of vector potentials that can result in the correct flowline field with different gauge functions. Here we are interested in only the meaningful ones. One way to investigate such vector potential is to utilize equation eq. 3.44, which can be easily applied to a simple problem such as a configuration with only Lambertian source and no optical devices (no reflection/no refraction). One can also evaluate vector potential using pure ray tracing results. Both are demonstrated in this book. In the case of ideal 2D nonimaging optics, specifically the class of Compound Elliptical Concentrators (CEC), we present here a way of deducing one of the vector potentials without eq. 3.44, using gauge invariance to solve the reflectance problem.

Figure 3.9 demonstrates the flowline in an asymmetric CEC configuration. $C1 - C2$ and $A1 - A2$ are the source and sink of vector field $\vec{D}$ (we can also reverse them). Although we can either demonstrate this as a concentrator ($A1 - A2$ being the light source) or illumination ($C1 - C2$ being the light source) problem, we will only discuss the case of $C1 - C2$ being the flowline source (and light source), the concentrator problem can be similarly solved. The mirrors are positioned along the outer flowlines given by the external thick lines. The flowlines (direction of $\vec{D}$) are the curved internal thin lines. The crossing lines $A1 - C2$ and $A2 - C1$ divide the flowlines into 4 regions. All the rays going through region I will directly arrive at $A1 - A2$ (i.e. without reflection of rays), region IV sees the light directly coming from the light source $C1 - C2$. Region II and region III see both rays directly from $C1 - C2$ and reflected rays of $C1 - C2$ from the mirror.

Because region I and region IV are the same configurations, we already know that the vector potential (or one of the many possible vector potential solutions) can be expressed as the difference between $PA1$ and $PA2$. Here for convenience, we can define the optical path length between point P and point $A2$ as L_{A2}, notice that L_{A2} is not a constant but a function that changes according to where P is, as shown in fig. 3.10. Using the result from [5] about vector potential (which will also be worked out in section 3.6), we can conclude that in region I the vector potential is $L_{A2} - L_{A1}$, similarly, in region IV the

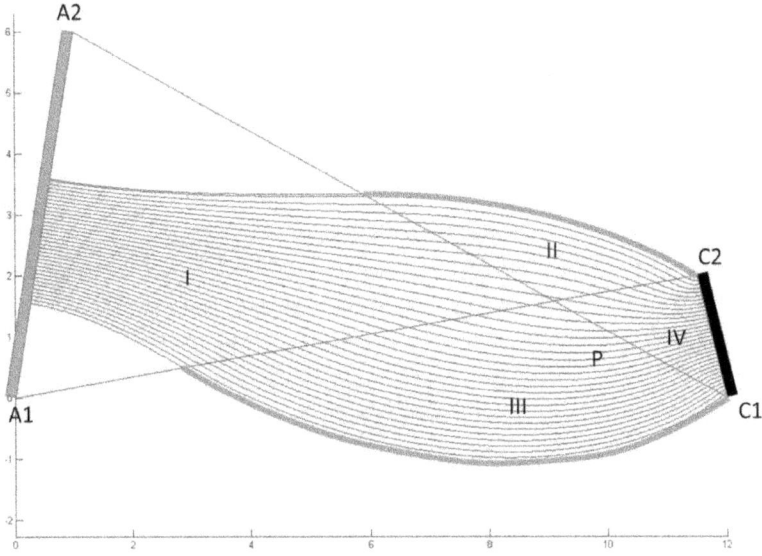

FIGURE 3.9: The configuration of an Asymmetric Compound Elliptical Concentrator (ACEC)

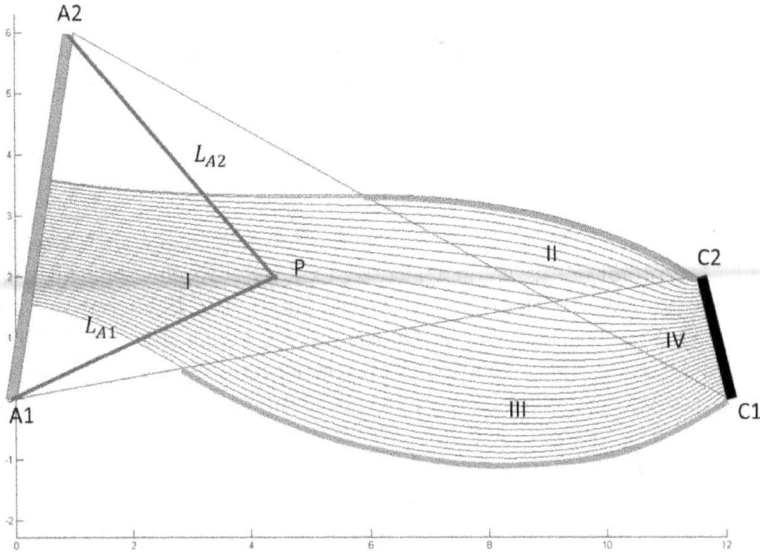

FIGURE 3.10: The vector potential in region I can be calculated by difference of string lengths between $P - A2$ and $P - A1$

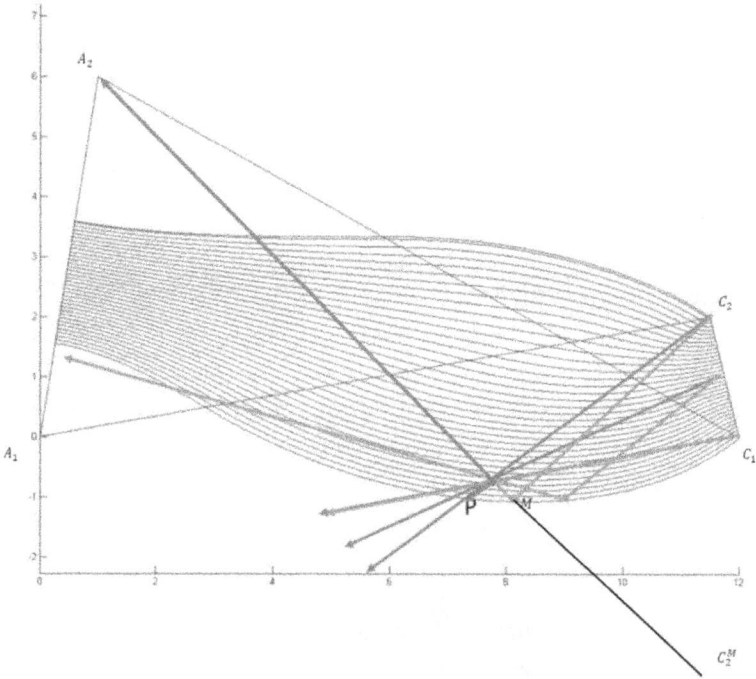

FIGURE 3.11: Vector potential field of region III

vector potential is $L_{C2} - L_{C1}$. Notice that we can add any scalar function with a divergence being 0 to this vector potential field ($\vec{A}$) without changing the vector field ($\vec{D}$), which includes, of course, any constant value. Then we can state for region I $L_{A2} - L_{A1} + \Phi_1$ and for region IV $L_{C2} - L_{C1} + \Phi_4$. Here Φ_1 and Φ_4 are the constant we add as the gauge, they can be functions, but in this case constant value is sufficient as we shall see later.

Next, we will work out the vector potential field in region II and region III, take any point P in region III as an example, we can scan all the rays (rays are represented with arrows) that originated from light source $C2 - C1$. As shown in fig. 3.11, we can start with ray $C2 \to P$, this ray will directly pass point P, so for the calculation of vector potential it needs to be included. As we rotate this ray clockwise, all the rays originating between $C2$ and $C1$ and directly pass point P will need to be included. If we further rotate the rays that pass P, we start to include the rays that will have one reflection at the bottom mirror. This will continue until the last ray $C2 \to M$, $M \to A2$ passes point P. In order to calculate vector potential, we need to identify the two edge points of the source, which become $C2$ and a reflected point C_2^M (notice that C_2^M moves as P moves). The path length of $C2 \to M \to P$ is the same as *constant*–$P \to A2$ due to the property of the elliptical mirror. If we rewrite the path length of $P \to A2$ as L_{A2}, the vector potential should be

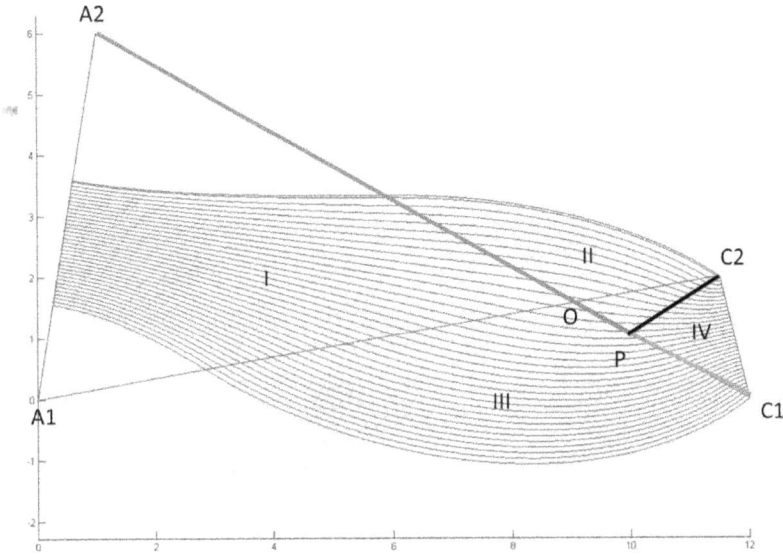

FIGURE 3.12: Boundary conditions

$L_{C2} - (constant - L_{A2})$, or $L_{C2} + L_{A2} - constant$. Similarly, we can work out the vector potential for region II as $-L_{A1} - L_{C1} + constant$, and we can write for region II $-L_{C1} - L_{A1} + \Phi_2$ and for region III $L_{A2} + L_{C2} + \Phi_3$

When P is on the boundary of region III and region IV, as shown in fig. 3.12, the equation

$$L_{C2} - L_{C1} + \Phi_4 = L_{A2} + L_{C2} + \Phi_3, \qquad (3.45)$$

has to be true as long as P remains on the line of $OC1$, we can rearrange the equation to be,

$$\Phi_4 - \Phi_3 = L_{A2} + L_{C1}, \qquad (3.46)$$

Similarly, we can derive the following equations

$$\Phi_2 - \Phi_1 = L_{A2} + L_{C1}, \quad P \in OA2 \qquad (3.47)$$

$$\Phi_1 - \Phi_3 = L_{A1} + L_{C2}, \quad P \in OA1 \qquad (3.48)$$

$$\Phi_2 - \Phi_4 = L_{A2} + L_{C1}, \quad P \in OC2 \qquad (3.49)$$

The overlapping position of all these four conditions is when P is at O. Notice that when P is at O, we can define L_1 as distance between $A1$ and $C2$, and L_2 as distance between $A2$ and $C1$, then the equations above reduce to (for P at O)

$$\Phi_4 - \Phi_3 = L_2, \qquad (3.50)$$

$$\Phi_2 - \Phi_1 = L_2 \tag{3.51}$$

$$\Phi_1 - \Phi_3 = L_1 \tag{3.52}$$

$$\Phi_2 - \Phi_4 = L_1 \tag{3.53}$$

which can be solved as

$$\Phi_1 = \frac{L_1 - L_2}{2}, \tag{3.54}$$

$$\Phi_2 = \frac{L_1 + L_2}{2} \tag{3.55}$$

$$\Phi_2 = \frac{-L_1 - L_2}{2} \tag{3.56}$$

$$\Phi_4 = \frac{-L_1 + L_2}{2} \tag{3.57}$$

and finally the vector potential in any region of the asymmetric CEC can be obtained as:

For region I

$$L_{A2} - L_{A1} + \frac{L_1 - L_2}{2}, \tag{3.58}$$

For region II

$$-L_{C1} - L_{A1} + \frac{L_1 + L_2}{2}, \tag{3.59}$$

For region III

$$L_{A2} + L_{C2} + \frac{-L_1 - L_2}{2}, \tag{3.60}$$

For region IV

$$L_{C2} - L_{C1} + \frac{-L_1 + L_2}{2}, \tag{3.61}$$

This demonstrates how gauge invariance works in solving the vector potential problem of asymmetric CEC configuration.

3.6   Some Irradiance Pattern Computation Examples

In this section we are going to study some examples of computation of $\vec{D}$ using the differential mathematical tools presented in previous sections. As the first examples of the proposed methodology we analyze simple symmetric systems. First we compute the irradiance produced at a point P in space by an infinite extruded Lambertian source , using the concept of vector potential (fig. 3.14): from eq. 3.42 the vector potential at P is

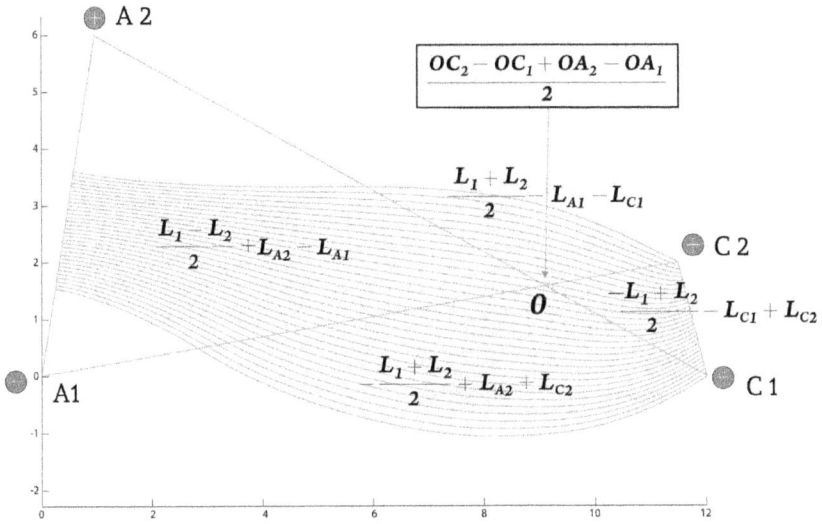

FIGURE 3.13: The solution for each region and the center point

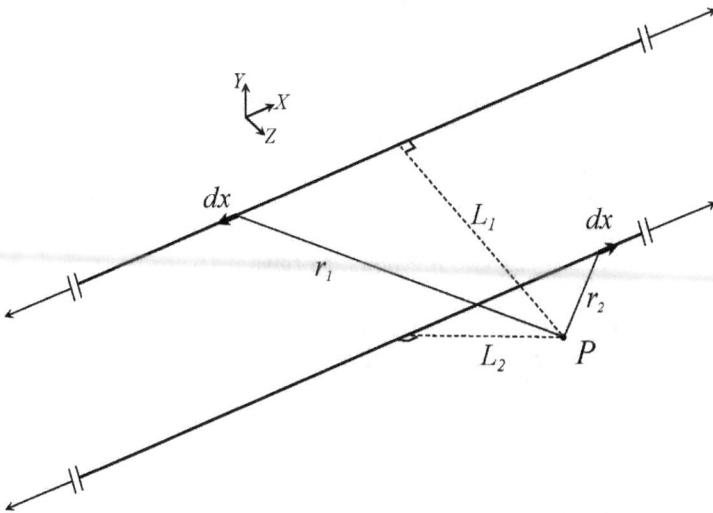

FIGURE 3.14: Sketch of the extruded source and parameters to compute the vector potential integral

$$A_x = \frac{B}{2} \left(\int_{-\infty}^{+\infty} \frac{1}{2} \ln (x^2 + L_2^2) dx - \int_{-\infty}^{+\infty} \frac{1}{2} \ln (x^2 + L_1^2) dx \right). \quad (3.62)$$

Note that for the configuration of fig. 3.14 the direction of $\vec{A}$ is at the X axis; the Y and Z components vanish. Using the properties of ln function and integrating

$$A_x = \frac{B}{4} \left[x \ln \left(\frac{x^2 + L_2^2}{x^2 + L_1^2} \right) + 2L_2 \arctan \left(\frac{x}{L_2} \right) - 2L_1 \arctan \left(\frac{x}{L_1} \right) \right]_{-\infty}^{+\infty}, \quad (3.63)$$

the limit of the first term in the summation of the right hand is

$$\lim_{x \to \infty} x \ln \left(\frac{x^2 + L_2^2}{x^2 + L_1^2} \right) = \lim_{x \to \infty} x \ln \left(1 + \frac{L_2^2 - L_1^2}{x^2 + L_1^2} \right) = \lim_{x \to \infty} x \left(\frac{L_2^2 - L_1^2}{x^2 + L_1^2} \right) = 0, \quad (3.64)$$

and finally we get

$$A_x = \frac{B}{2} \left(L_2 \left(\frac{\pi}{2} + \frac{\pi}{2} \right) - L_2 \left(\frac{\pi}{2} + \frac{\pi}{2} \right) \right) = \frac{B\pi}{2} (L_2 - L_1). \quad (3.65)$$

Then the component D_z at point P, which is the irradiance incident on a plane detector parallel to the XY plane, is obtained as

$$D_z = \frac{\partial A_x}{\partial y} = \frac{B\pi}{2} \frac{d}{dy} (L_2 - L_1). \quad (3.66)$$

Figure 3.15 shows the irradiance profile for an extruded Lambertian source of 20 mm length and a detector at $z = 50 \, mm$ from the source, computed using eq. 3.66, compared with the irradiance obtained by raytracing; it shows great agreement between the two methods. Note that the ordinate axis represents irradiance in arbitrary units (a.u.), which means that no calibration is required; only the normalized pattern has been analyzed. We will use this kind of representation of irradiance patterns through this book to avoid the need to calibrate irradiance units in the comparisons between theoretical results and raytracing checks , for simplicity.

As the second example we are going to study the vector potential for a rotational symmetric Lambertian source , a Lambertian disk, fig. 3.16. To do that we use cylindrical coordinates (ρ, θ, z). Let $\vec{A} = (A_\rho, A_\theta, A_z)$; assume that $A_\theta \neq 0$, and to obtain $\vec{A}$ we use Stoke's theorem:

$$\oint_l \vec{A} \cdot \vec{dl} = \int_s \vec{D} \cdot \vec{ds}. \quad (3.67)$$

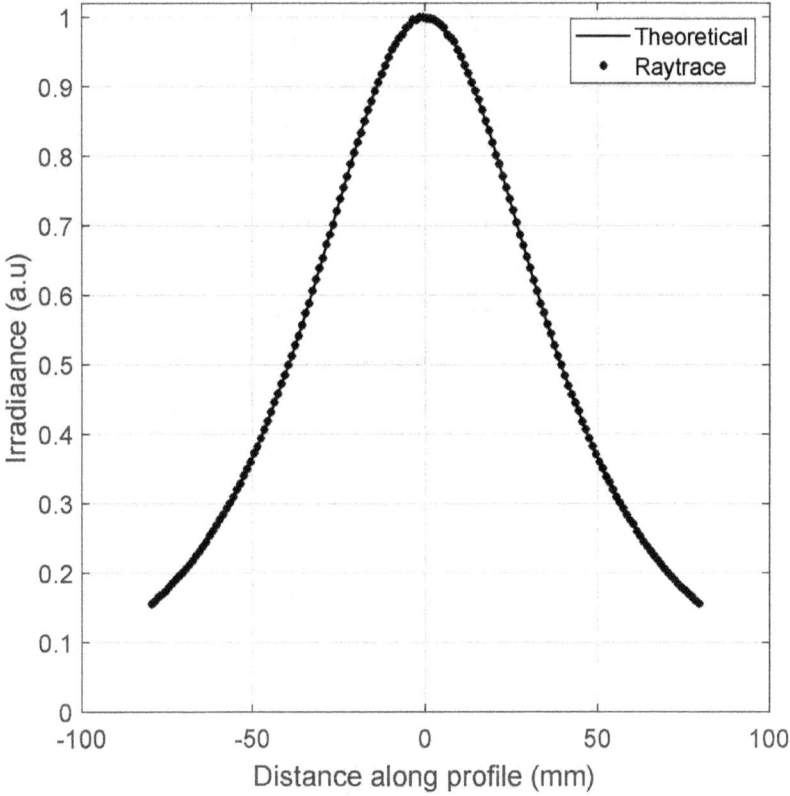

FIGURE 3.15: Comparison of irradiance profiles obtained from $\vec{D}$ computation using eq. 3.66 and from raytracing, for a 2D system

Let the source of $\vec{D}$ be a disk of radius f in plane $z = 0$. Let's evaluate $\vec{A}$ at $z = 0$: considering that $|\vec{D}|_{z=0} = \pi$ due to the Lambertian surface, the flux through the disk of radius a is

$$\oint_{z=0} \vec{A} \cdot \vec{dl} = 2\pi a A_\theta = \pi^2 a^2, \tag{3.68}$$

thus

$$A_{\theta(z=0)} = \frac{\pi a}{2}; \tag{3.69}$$

considering a flux tube, the line integral of $\vec{A}$ does not change along a flux tube due to flux conservation; then for any other plane $z \neq 0$

$$\oint \vec{A} \cdot \vec{dl} = 2\pi \rho A_\theta = \pi^2 a^2, \tag{3.70}$$

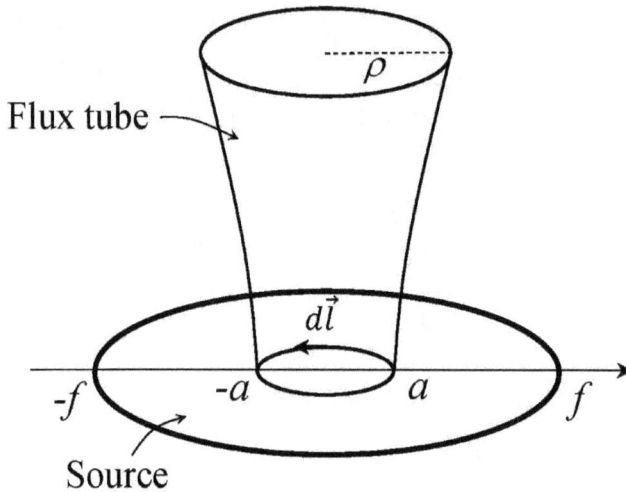

FIGURE 3.16: Parameters to compute vector potential integral for
rotational source

therefore

$$A_{\theta(z=0)} = \frac{\pi a^2}{2\rho};$$
(3.71)

to conclude that

$$\vec{A} = (A_\rho, A_\theta, A_z) = \left(0, \frac{\pi a^2}{2\rho}, 0\right),$$
(3.72)

where $A_\theta\rho$ =constant, acting like the conservation momentum along the axis.
Using the property of a hyperbola of conserving the difference of a point's
distance to the two foci, fig. 3.17, the a can be found as

$$a = \frac{1}{2}\left(\sqrt{(\rho+f)^2 + z^2} - \sqrt{(\rho-f)^2 + z^2}\right).$$
(3.73)

OZ is the optical axis, and the irradiance at a plane $z = const.$ can be computed as

$$D_z = \frac{1}{\rho}\frac{\partial(\rho A_\theta)}{\partial\rho},$$
(3.74)

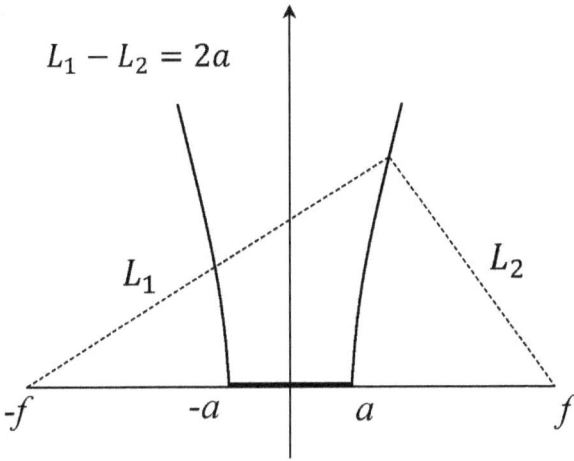

FIGURE 3.17: Property of a hyperbola flux tube

$$D_z = \cfrac{\pi \left(\sqrt{(\rho + f)^2 + z^2} - \sqrt{(\rho - f)^2 + z^2} \right)}{4\rho \left[\cfrac{(\rho + f)}{\sqrt{(\rho + f)^2 + z^2}} - \cfrac{(\rho - f)}{\sqrt{(\rho - f)^2 + z^2}} \right]}. \tag{3.75}$$

We have checked this result for a circular Lambertian source of 10 mm radius and a $160 \times 160\,mm^2$ detector placed 50 mm from the source; fig. 3.18 shows the irradiance profile computed by eq. 3.75 compared with a raytracing computation.

As a third example of irradiance vector computation, we will study some simple examples of non-symmetric optical systems. We will start with free propagation of radiation from square and rectangular sources. The sources lie in XY plane, and OZ will be the optical axis. In these examples we have computed the irradiance map orthogonal to the optical axis for different configurations of sources and detectors. Figure 3.19 shows the sketch for the optical configuration to be analyzed. The irradiance E at detector can be computed

Field Theory Elements

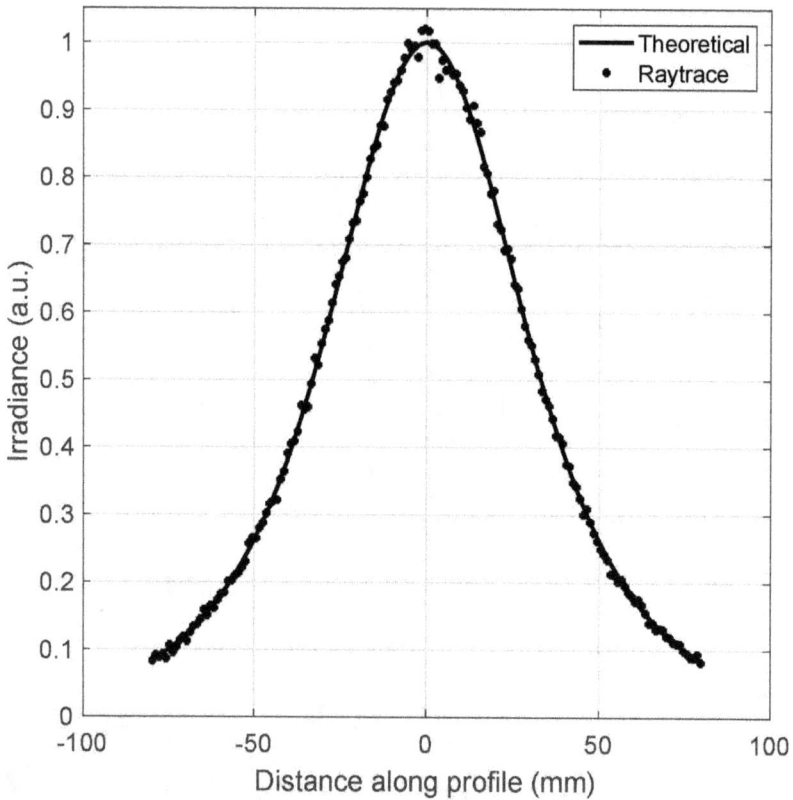

FIGURE 3.18: Irradiance profile comparison at distance $z = 50\,mm$ from the
circular source of radius $10\,mm$, computed using eq. 3.75 and raytracing

in different ways, first using vector potential

$$E = D_z = \frac{\partial A_y}{\partial x} - \frac{\partial A_x}{\partial y}, \tag{3.76}$$

and computing A_y and A_x using eq. 3.42 . The second way to compute the
irradiance E at the detector is by contour integral , using eq. 3.19, which
avoids a derivation in the process. The third way to compute D_z is by means of
eq. 3.32, which for square or rectangular sources is the easiest way to compute
it. Of course the three computation methods of D_z produce the same results.
As an example, we have done the computations for different configurations
of source detector systems. The comparisons between $\vec{D}$ computations and
raytracing are shown in figs. 3.20, 3.21 and 3.22.

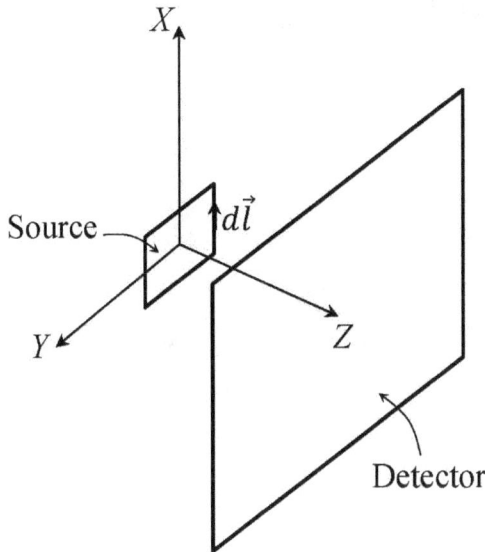

FIGURE 3.19: Sketch of the 3D analyzed system of a non-symmetric source

Figure 3.20 shows the irradiance map and irradiance profiles produced by a Lambertian square source of $160 \times 160 \, mm^2$ and a detector of $200 \times 200 \, mm^2$ placed 20 mm in front of the source, fig. 3.21 shows the irradiance map and irradiance profiles produced by a Lambertian rectangular source of $20 \times 180 \, mm^2$ and a detector of $200 \times 200 \, mm^2$ placed 5 mm in front of the source and fig. 3.22 shows the irradiance map and irradiance profiles produced by a Lambertian rectangular source of $20 \times 180 \, mm^2$ and a detector of $200 \times 200 \, mm^2$ placed 50 mm in front of the source. The subfigures a) show $\vec{D}$ computations and subfigures b) show raytracing results. All configurations show great agreement between $\vec{D}$ computation and raytracing. There are no significant differences between the proposed methods of computing $\vec{D}$.

3.7 The Curl of $\vec{D}$ and Quasipotential Fields

By differentiation of eq 3.19, it is possible to obtain an expression for the curl of $\vec{D}$:

$$\nabla \times \vec{D} = \frac{B}{2} \oint_l \frac{\vec{dl}}{r^2}. \tag{3.77}$$

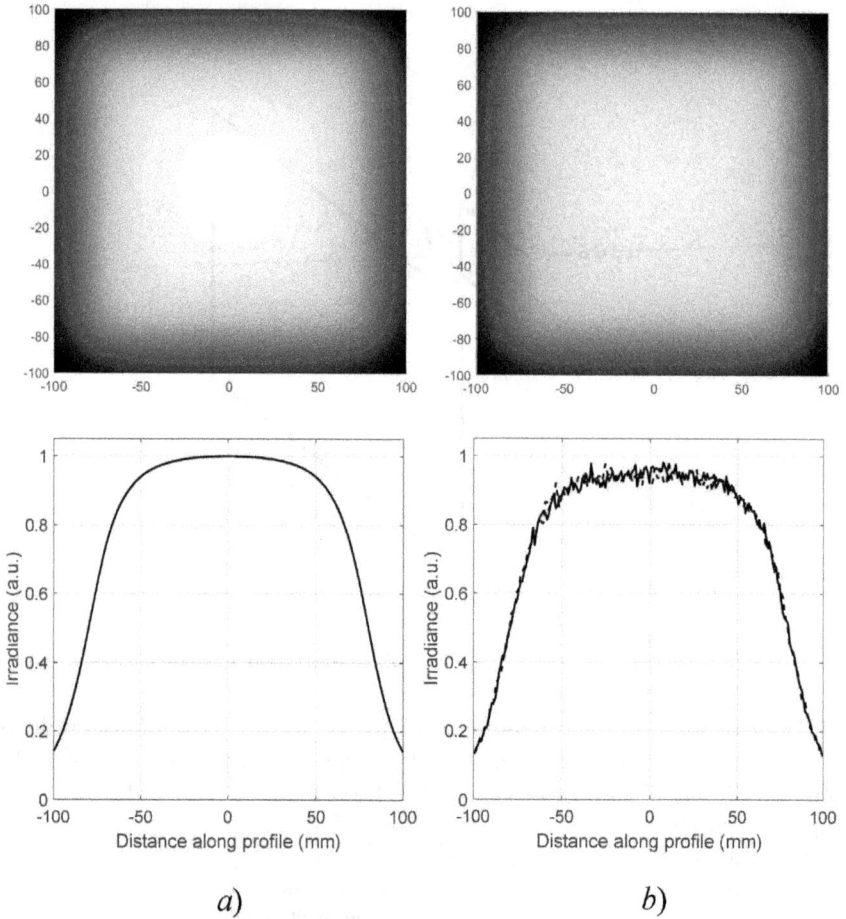

FIGURE 3.20: Irradiance map and profile for a square source of 160×160 mm^2, detector of 200×200 mm^2, Z distance 20 mm, a) $\vec{D}$ computation using vector potential eq. 3.42 or contour integration eq. 3.19 or eq. 3.32, b) raytracing

In general this contour integral is nonvanishing. Only a few high symmetric $\vec{D}$ fields are irrotational fields: the field produced by a perfect spherical source or a perfect infinite plane source [42]. For these fields we can define a scalar potential ϕ, and equipotential surfaces are orthogonal surfaces to $\vec{D}$. Nevertheless, for most $\vec{D}$ it is possible to define a scalar integration function $\mu(x, y, z)$ such that $\mu\vec{D}$ is irrotational [43]. If this integration function exists, there exists another function Φ_q:

$$\mu\vec{D} = \nabla\Phi_q. \tag{3.78}$$

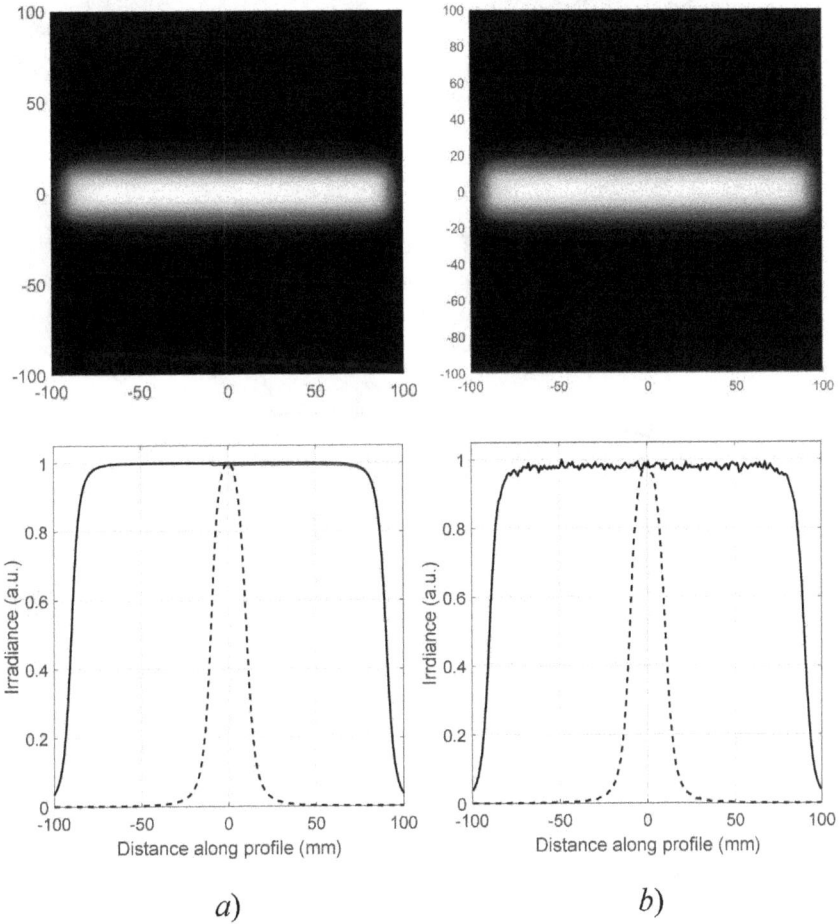

FIGURE 3.21: Irradiance map and profile for a rectangular source of $20 \times 180 \, \text{mm}^2$, detector of $200 \times 200 \, \text{mm}^2$, Z distance 5 mm, a) $\vec{D}$ computation using vector potential eq. 3.42 or contour integration eq. 3.19 or eq. 3.32, b) raytracing

Taking the curl to both sides of the equation, we have

$$\mu \nabla \times \vec{D} + \nabla \mu \times \vec{D} = 0. \tag{3.79}$$

The scalar product of this equation with $\vec{D}$ is

$$\mu \vec{D} \cdot \nabla \times \vec{D} = 0, \tag{3.80}$$

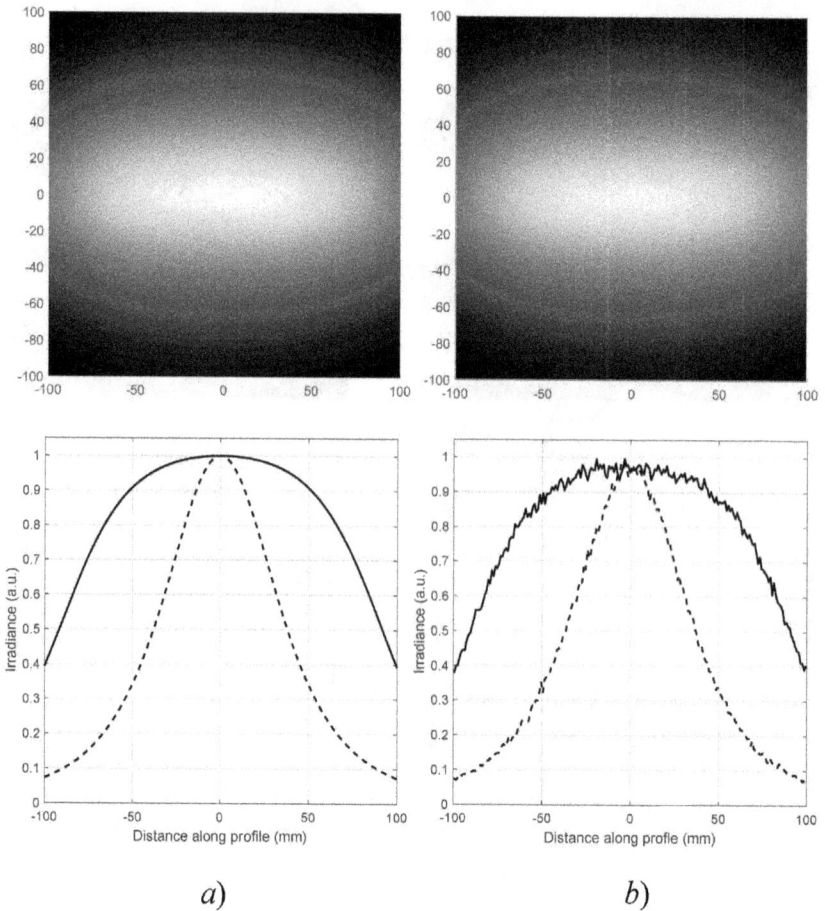

FIGURE 3.22: Irradiance map and profile a for rectangular source of
$20 \times 180\,\text{mm}^2$, detector of $200 \times 200\,\text{mm}^2$, Z distance 50 mm, a) $\vec{D}$
computation using vector potential eq. 3.42 or contour integration eq. 3.19
or eq. 3.32, b) raytracing

and as μ is not zero,

$$\vec{D} \cdot \nabla \times \vec{D} = 0. \tag{3.81}$$

Equation 3.81 is the necessary and sufficient condition for the existence of a
integrating factor μ , which reduces to unity in the special case of $\nabla \times \vec{D} = 0$.
Φ_q is the quasipotential function . Using the properties of the gradient, the
surfaces $\Phi_q = const$, being everywhere perpendicular to the vector $\mu\vec{D}$, are
also perpendicular to $\vec{D}$ and are called quasipotential surfaces.

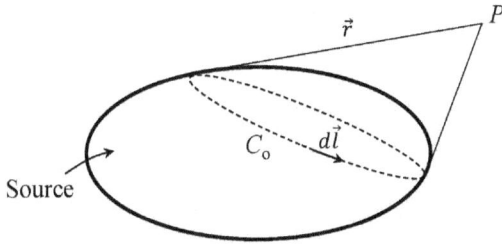

FIGURE 3.23: Contour integral representation of $\vec{D}$

The eq. 3.81 is satisfied for most cases of practical interest, including fields with axial symmetry, plane fields, fields produced by uniform sources that conform in shape to a coordinate surface of an orthogonal coordinate system or fields whose defining contour curves C_o are planar. This last means that the field is produced by a source for which the contour C_o of the integral of eq. 3.19 is planar at any point in space, fig. 3.23. This can be shown by taking the scalar product of $\vec{D}$ and its curl from eq. 3.19 and eq. 3.77, apart from constant factors

$$\vec{D} \cdot \nabla \times \vec{D} = \oint_l \int_l \frac{\vec{dl} \times \vec{r} \cdot \vec{dl'}}{r^2 r'^2}, \tag{3.82}$$

where l and l' denote the position of any two points on contour C_0, and $\vec{r}$ and $\vec{r'}$ denote the position vector of any point P in the space, from any two points of contour C_0. By interchanging the variables of integration l and l' we have

$$\vec{D} \cdot \nabla \times \vec{D} = \oint_l \int_l \frac{\vec{dl'} \times \vec{r'} \cdot \vec{dl}}{r^2 r'^2}, \tag{3.83}$$

using the rule of triple product in the integrand and adding both equations, we obtain

$$\vec{D} \cdot \nabla \times \vec{D} = \frac{1}{2} \oint_l \int_l \frac{\vec{dl'} \times \vec{dl} \cdot (\vec{r} - \vec{r'})}{r^2 r'^2}, \tag{3.84}$$

where $\vec{r} - \vec{r'}$ is the vector connecting the elements $\vec{dl'}$ and $\vec{dl}$. Therefore if contour C_0 is a plane curve, the integrand in eq. 3.84 vanishes by the orthogonality between vectors in the numerator, and $\vec{D}$ satisfies the eq. 3.81, which means that when the contour C_o is a plane a quasipotential exists.

For example, Moon [5] studied quasipotential fields using orthogonal coordinate systems. From oblate-spheroidal coordinates (η, ϕ, ψ), fig. 2.16, he deduced the quasipotential produced by a uniform circular disk source

$$\Phi_q = \frac{2}{\pi} \Phi_0 \cot^{-1}(\sinh \eta). \tag{3.85}$$

The surfaces of $\Phi_q = const$ produce orthogonal surfaces to $\vec{D}$, in this case corresponding to the surfaces of $\eta = const$ as expected, being the integrating factor

$$\mu = -\frac{C}{aB\sqrt{(1 - \sin^2 \phi)}}, \tag{3.86}$$

where a is the radius of the disk, B is the brightness of the source and C can be obtained by boundary conditions. Other quasipotential fields were analyzed using other orthogonal coordinate systems [5].

In the examples used in this book we can always construct orthogonal surfaces to $\vec{D}$, including those obtained by the conformal mapping of section 2.4. These orthogonal surfaces have interesting properties for optical design techniques, and in the next chapter we will analyze the role of these surfaces orthogonal to $\vec{D}$ in the design of refractive optical elements. Now it is interesting to note a property that arises for symmetric systems. In section 1.6 we stated that for a Lambertian strip source, the field lines are confocal hyperbolas, and along any field line the difference in the optical path length of edge rays is constant. Then, following this, the lines orthogonal to field lines have the following interesting property: the sum of the optical path length of each edge ray is constant. This property provides us a tool to compute orthogonal surfaces, mainly for symmetric axis optical systems, fig. 3.24.

3.8   Basic Introduction to Lorentz Geometry

Lorentz geometry is an exciting field of mathematical research that can be seen as part of differential geometry as well as mathematical physics. It represents the mathematical foundation of the general theory of relativity, and the rapid development of Lorentz geometry was due to Einstein's general relativity theory of gravitation. Lorentz geometry was first applied to nonimaging optics by Gutierrez [36] [44]. We will briefly discuss Lorentz geometry here, but the interested reader can consult [45] for some background on differential geometry. The main property of Lorentzian vector space is that the definition of scalar product is different than in usual Euclidean vector space. Let there be two vectors in $\mathbb{R}^3$, expressed in the canonical basis $\vec{u} = (u_1, u_2, u_3)$ and $\vec{v} = (v_1, v_2, v_3)$. The Lorentzian scalar product is defined as

$$\langle \vec{u} \cdot \vec{v} \rangle_L = u_1 v_1 + u_2 v_2 - u_3 v_3, \tag{3.87}$$

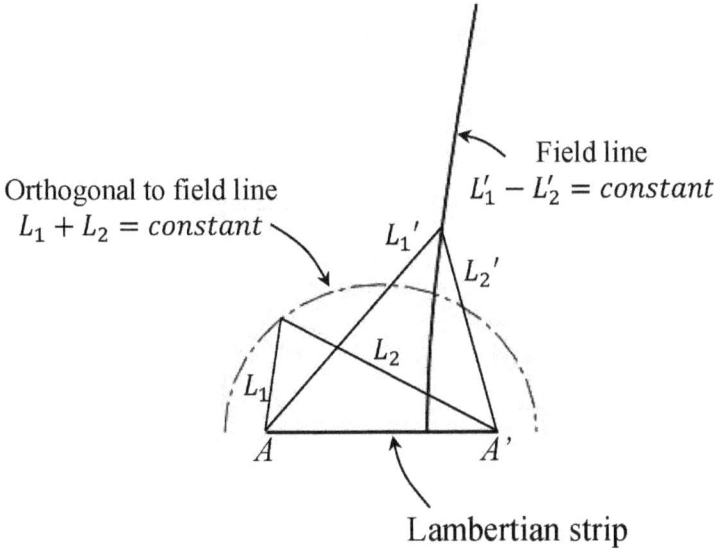

Orthogonal to field line
$L_1 + L_2 = constant$

Field line
$L_1' - L_2' = constant$

L_1'

L_2'

L_2

L_1

A A'

Lambertian strip

FIGURE 3.24: Properties of field lines and orthogonal lines

where the subindex L means the Lorentzian scalar product, with the associated signature $(+, +, -)$. From this definition of scalar product it is possible to establish what is named the causal classification of vectors:

1. $\vec{u}$ is a timelike vector if $\langle \vec{u} \cdot \vec{u} \rangle_L < 0$.

2. $\vec{u}$ is a lightlike vector if $\langle \vec{u} \cdot \vec{u} \rangle_L = 0$ and $\vec{u} \neq 0$.

3. $\vec{u}$ is a spacelike vector if $\vec{u} = 0$ or $\langle \vec{u} \cdot \vec{u} \rangle_L > 0$.

For our purposes we are interested in lightlike vectors . In Lorentz vector space the lightlike vectors are in a cone surface, fig. 3.25. This property suggests the use of Lorentz geometry as the mathematical tool to analyze the cone of edge rays introduced in section 1.5, which are basic elements in nonimaging optical design.

Gutierrez et al. [36] applied Lorentz geometry, as a branch of differential geometry, to nonimaging optics. They provided the matrix G, produced by a source/receiver with rotational symmetry, which is the Gram matrix of the Lorentz metric in the canonical basis of $\mathbb{R}^3$, in a way that the eigenvectors of G provide the direction of $\vec{D}$ and the eigenvalues of G provide the modulus of $\vec{D}$,

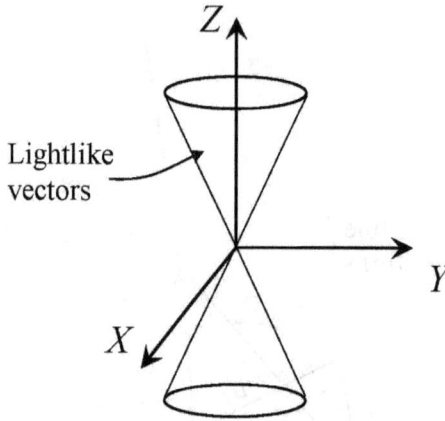

FIGURE 3.25: Lightlike vectors in Lorentz space

at any point in space P. The matrix G must fulfill the following constraints:

1. The eigenvectors of G at any point P produce an orthogonal basis B_P.

2. The lightlike vectors of the new basis are edge ray vectors of the source/receiver.

3. The trajectories of lightlike vectors in the new basis are straight lines.

In this and the following section we are going to study the fundamentals and properties of this matrix G. Let $B = [e_1, e_2 e_3]$, fig. 3.27, be the canonical basis vector field of an open subset M of the $\mathbb{R}^3$, which is orthogonal with respect to the usual Euclidean metric. For each point P on M, B_P is the canonical basis of the tangent space $T_P M$. A Lorentz metric g on M is a map that assigns to every point P of M a bilinear map $g{:}T_P M \times T_P M \to \mathbb{R}$, with some aditional properties. These properties may be expressed through the matrix G associated with the map g in the basis B. Using this matrix, one can formulate the map as $g_P(Y, Y') = Y^t G(P) Y'$, where Y and Y' are two vectors of $T_P M$ (the superscript t denotes transposition). The matrix G must have the following three properties:

1. G is symmetric on every point P of M.

2. $|G| \neq 0$ on every point of M.

3. G has one negative and two positive eigenvalues on every point of M. This is expressed by saying that g has the signature $(+, +, -)$.

The elements of G, g_{ij} are functions based on M; if they are differentiable of class C^∞ the above properties characterize a Lorentz metric based on M. The set of all lightlike vectors forms the cone of edge rays, whose equation can be written as $Y^t G Y = 0$.

3.9   Application of Lorentz Geometry to the Evaluation of $\vec{D}$

As we have mentioned in the previous section, M. Gutierrez et al. [36] applied Lorentz geometry to the study of $\vec{D}$. They showed that using Lorentz formalism, as a branch of differential geometry, it is possible to obtain the field lines and flux tubes of $\vec{D}$. In this section we show that it is also possible to obtain the irradiance pattern produced by Lambertian sources using Lorentz geometry formalism [46]. Lorentz geometry provides a cone structure on an open subset M of the $\mathbb{R}^3$; that is, a cone at each point P of M. This cone is called the light cone in Einstein's theory. The way to obtain the cone structure at any point in space is through the matrix G defined in the previous section and provided by M. Gutierrez et al. [36]. In nonimaging optics the light cone is what we have called the cone of edge rays in section 1.5.

Let us look at the role of the eigenvalues and eigenvectors of the matrix G of a Lorentz metric g. Because G is a symmetric matrix, it can be diagonalized, leading to a matrix $G' = \operatorname{diag}(\lambda_1, \lambda_2, \lambda_3)$, where $\lambda_1, \lambda_2, \lambda_3 \in \mathbb{R}$. This transformation from G to G' is just a change of basis, and thus G' may be expressed as $G' = A^{-1} G A$, where A is the matrix changing between a basis B'_P which is orthogonal according to canonical basis B_P, which is also orthogonal. The new basis is formed by three eigenvectors of G. The columns of A are the components of these eigenvectors in the canonical basis B_x, and the elements $\lambda_1, \lambda_2, \lambda_3$ of the diagonal G' are the eigenvalues of G associated with the corresponding eigenvector. Because g is Lorentzian, the signature is $(+, +, -)$; thus only one eigenvalue is negative and the other two are positive. There is no loss of generality if we assume that $\lambda_3 < 0$ and $|A| = 1$. Thus the basis B'_P in which G' is a diagonal matrix has a nice geometric interpretation, since it can be obtained from the canonical basis B_P by rotation.

Now the study of the cone structure in the basis B'_P produces three unit eigenvectors $B'_P = (\vec{u}_W, \vec{u}_V, \vec{u}_D)$ of G associated with eigenvalues λ_1, λ_2 and λ_3 respectively. Let $Y \neq 0$ be a lightlike vector. If the components of Y in the basis B'_P are (y_1, y_2, y_3), the equation of the lightlike vectors in the basis is $Y^t G' Y = 0$, that is

$$\lambda_1 y_1^2 + \lambda_2 y_2^2 + \lambda_3 y_3^2 = 0, \tag{3.88}$$

New basis B'_P

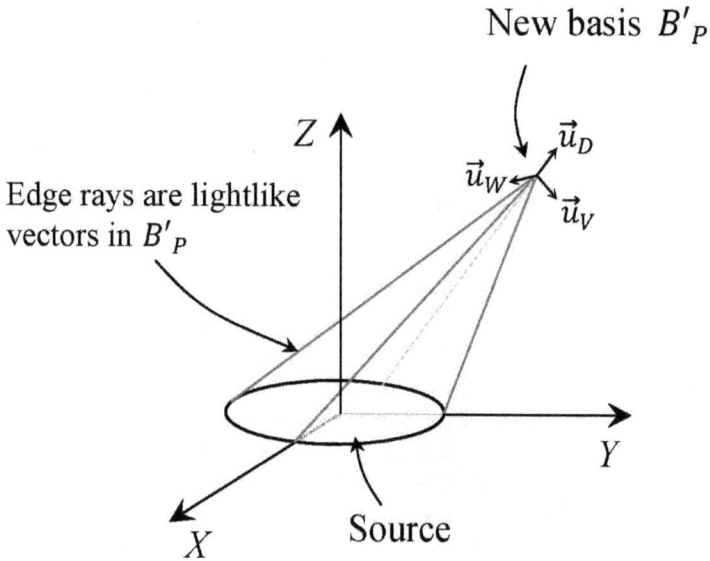

FIGURE 3.26: Change of basis to B'_P, note that this figure is equal to fig. 3.4

which is a cone whose axis is parallel to the unit vector $\vec{u}_D$ (remember that $\lambda_3 < 0$ and $\lambda_1, \lambda_2 > 0$). The intersection of the cone and the plane $y_3 = 1$ is the ellipse

$$\lambda_1 y_2^2 + \lambda_2 y_2^2 = -\lambda_3. \tag{3.89}$$

The ellipse axes are parallel to the unit vectors $\vec{u}_W$ and $\vec{u}_V$ (see fig. 3.26), and their respective semiaxis lengths are $(-\lambda_3/\lambda_1)^{1/2}$ and $(-\lambda_3/\lambda_2)^{1/2}$. To summarize:

1. The cone of edge rays at point P is elliptic, $\forall P \in M$.

2. The direction of the elliptic cone axis is the direction of $\vec{D}$, with $\vec{u}_D$ being the eigenvector of G associated with the negative eigenvalue of G.

3. The directions of the principal axes of the elliptic cone are the directions of $\vec{u}_W$ and $\vec{u}_V$, the eigenvectors of G associated with the positive eigenvalues of G.

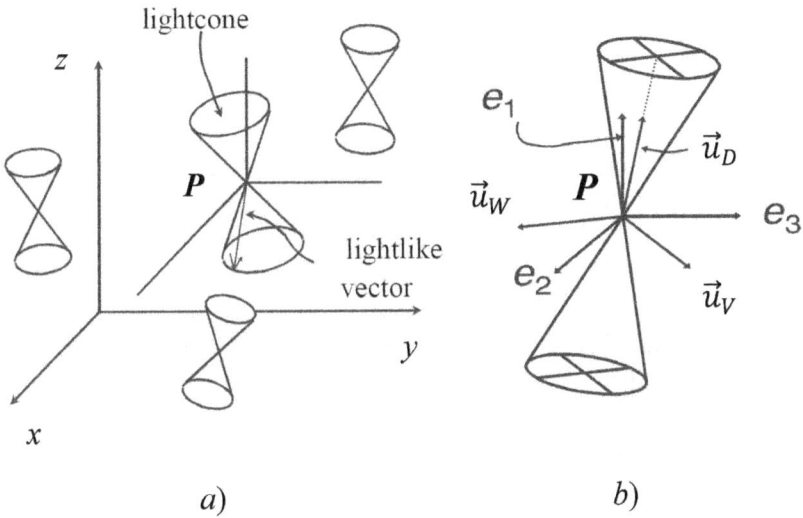

FIGURE 3.27: a) Lorentz geometry provides a cone structure in M, formed by a set of lightlike vectors at each point. b) The vectors $(\vec{u}_D, \vec{u}_W, \vec{u}_V)$ define the three planes of symmetry of the cone of edge rays at the point P

Let us consider a medium with a constant refractive index $n = 1$. Thus the ray trajectories will be straight lines. In the Lorentz geometry approach of Gutierrez et al. [36], a restricted optical condition was used, which states that every lightlike geodesic of the Lorentz metric must be a geodesic of the Euclidean metric. The application of this condition forces the lightlike curves to be geodesic of both the Lorentz metric and the Euclidean metric. This restriction allows us to formulate the optical condition through a system of differential equations. It is possible to express the restricted optical condition as follows: If γ is a C^2 curve in M, a open set of $\mathbb{R}^3$ such that $\dot{\gamma}(t) \neq 0$ and $\forall t \in \mathbb{R}$, then for $k \in \{1, 2, 3\}$

$$\left. \begin{array}{c} \ddot{\gamma}^k(t) + [\dot{\gamma}(t)]^k \Gamma^k[\gamma(t)]\dot{\gamma}(t) = 0 \\ \dot{\gamma}^k(t) G[\gamma(t)]\dot{\gamma}(t) = 0 \end{array} \right\} \Rightarrow \ddot{\gamma}^k(t) = 0 \qquad (3.90)$$

The upper-left equation established that γ is a geodesic curve in Lorentz geometry, the lower-left implies that γ is a lightlike curve, and the equation on the right-hand side is the condition for γ to be a Euclidean geodesic curve.

Γ^k is a matrix formed by the elements Γ^k_{ij}, which are the Christoffel symbols of the Lorentz metric [45]. The expression of these symbols is

$$\Gamma^k_{ij} = \frac{1}{2} \sum_{m=1}^{3} g^{km}(g_{jm,i} + g_{mi,j} - g_{ij,m}), \qquad (3.91)$$

where $G = (g_{ij})$, and the subscript after the comma denotes the partial derivative $g_{jm,i} = \frac{\partial g_{jm}}{\partial x_i}$. The eq. 3.90 is fulfilled if

$$\Gamma^k = f^k G, \; k \in \{1, 2, 3\}, \qquad (3.92)$$

where f^k are arbitrary functions on M. Equation 3.92 will be called restricted optical equations. Using the expressions for the Christoffel symbols Γ^k_{ij} we can write the restricted optical equations

$$\frac{\partial g_{im}}{\partial x_j} = \sum_{k=1}^{3}(g_{mk}f^k g_{ij} + g_{ik}f^k g_{jm}), \; i, j, m \in \{1, 2, 3\}, \qquad (3.93)$$

which is a system of first order differential equations, of which the solutions are g_{ij}, the components of G Lorentz matrix which we are looking for. Because G is symmetric it is possible to reduce the number of differential equations in the systems. Gutierrez [36] solved this system of equations providing the inverse matrix G^{-1}:

$$G^{-1} = \begin{pmatrix} -ax^2 - 2b_2 x + m & n - b_3 x - b_2 y - axy & k - b_1 x - b_2 z - axz \\ n - b_3 x - b_2 y - axy & -ay^2 - 2b_3 y + p & l - b_1 y - b_3 z - ayz \\ k - b_1 x - b_2 z - axz & l - b_1 y - b_3 z - ayz & -az^2 - 2b_1 z + d \end{pmatrix},$$
$$(3.94)$$

where $a, b_1, b_2, b_3, d, k, l, m, n, p \in \mathbb{R}$ are the 10 parameters obtained from the integration of the system of differential equations 3.93. For simplicity it is useful to start with systems with rotational symmetry. Gutierrez [36] showed that rotational symmetry is fulfilled if $b_2 = b_3 = k = l = n = 0$ and $p = m$ and that the cone of edge rays are tangent to the surface $|G^{-1}| = 0$, and he provides a classification of surfaces $|G^{-1}| = 0$ which can be considered Lambertian sources of radiation. Summarizing the set of parameters provides the geometrical configuration of the source which produces $\vec{D}$. As the first example we are going to analyze a circular disk source with rotational symmetry about the Z axis: we can rename b_1 as b; the plane of the circular disk source is $z = \frac{b}{a}$; and choosing $z = 0$ implies $b = 0$. Assume a circular disk source, $ad - b^2 = 0$, $a > 0$ and $d = 0$. The radius of the disk is $r = \left(\frac{m}{a}\right)^{1/2}$. At the position $a = 1$, $m = r^2$. For this set of parameters the G matrix becomes

$$G = \frac{1}{|G^{-1}|}\begin{pmatrix} -mz^2 & 0 & mzx \\ 0 & -mz^2 & mzy \\ mzx & mzy & m^2 - (x^2 + y^2)m \end{pmatrix}, \qquad (3.95)$$

being $|G^{-1}| = -m^2 z^2$. For a circular disk it is well known that the field lines of $\vec{D}$ are hyperbolas with their focal point at the edge of the disk. Taking into account that $\vec{u}_D$ is the eigenvector associated with a negative eigenvalue, we are going to prove that tangent vectors to hyperbolas are eigenvectors of matrix 3.95 associated with a negative eigenvalue, . Let there be a disk of radius 1, and consider only the plane XZ; making $y = 0$ in eq. 3.95 we have the 2×2 matrix

$$G = \frac{1}{|G^{-1}|} \begin{pmatrix} -z^2 & zx \\ zx & 1 - x^2 \end{pmatrix}. \tag{3.96}$$

Let the equation of the hyperbola be

$$\frac{x^2}{\beta^2} - \frac{z^2}{\alpha^2} = 1, \tag{3.97}$$

where $\alpha^2 + \beta^2 = 1$ and the focal points are located at $(\pm 1, 0)$. The parametric equation of the hyperbola can be written as

$$x = \beta \cosh(t), \quad z = \alpha \sinh(t), \quad t \in (0, 2\pi) \tag{3.98}$$

and its tangent

$$x' = \beta \sinh(t), \quad z' = \alpha \cosh(t), \quad t \in (0, 2\pi). \tag{3.99}$$

We need to prove that the tangent vector $\gamma'(t)$ is an eigenvector of eq. 3.96, written with the parametrization

$$G\gamma'(t) = -\frac{1}{\alpha^2 \sinh^2(t)} \begin{pmatrix} -\alpha^2 \sinh^2(t) & \alpha\beta \cosh(t) \sinh(t) \\ \alpha\beta \cosh(t) \sinh(t) & 1 - \beta^2 \cosh^2(t) \end{pmatrix} \begin{pmatrix} \beta \sinh(t) \\ \alpha \cosh(t) \end{pmatrix} \tag{3.100}$$

and

$$G\gamma'(t) = -\frac{1}{\sinh^2(t)} \begin{pmatrix} \beta \sinh(t) \\ \alpha \cosh(t) \end{pmatrix}, \tag{3.101}$$

where the eigenvalue $-\frac{1}{\sinh^2(t)} < 0$. Then the tangent vector $\gamma'(t)$ is proportional to $\vec{u}_D$, the unit vector is in the direction of $\vec{D}$ and the field lines are hyperbolas, as expected.

Once we have the direction of $\vec{D}$ at any point in space, as the eigenvector associated with the negative eigenvalue of matrix 3.95, the next step in our analysis is to obtain the modulus of $\vec{D}$. For rotational symmetric systems at any point in space, P, the cone of edge rays has an elliptic base. Equation 3.89 provides the semiaxis of the ellipse at a distance 1 to the point P in the $\vec{u}_D$ direction. It is easy to obtain the angles of the elliptic cone of edge rays, and using eq. 3.9, $D = B \sin \theta \sin \varphi$, we can compute the modulus of $\vec{D}$ at any

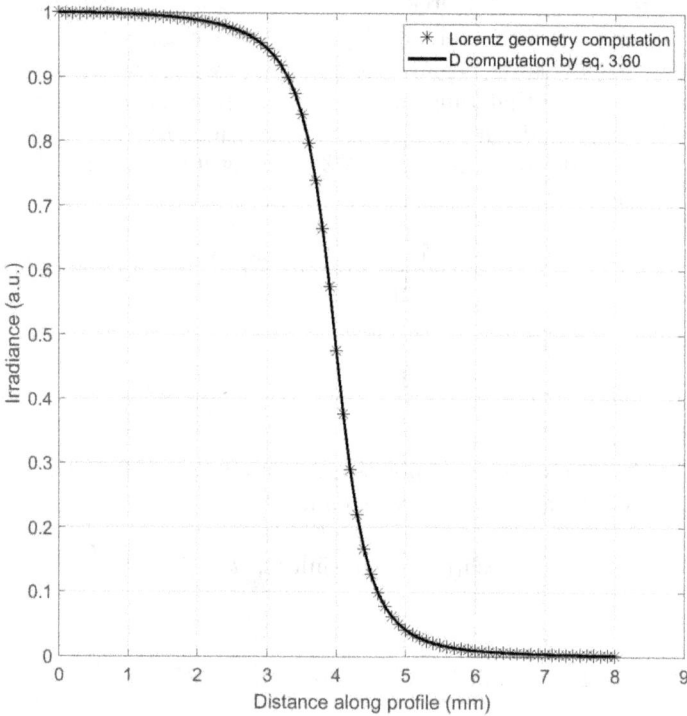

FIGURE 3.28: Comparison of irradiance obtained at $z = 2.5\,mm$ for a circular disk of radius $r = 4\,mm$ between the Lorentz geometry method and eq. 3.75

point P. We have done numerical computations of $\vec{D}$, using Lorentz geometry, for different radii of the disk and at different planes for detection, with $z = $ constant; and we have compared the results with Lorentz geometry with results obtained with eq. 3.75. Figure 3.28 shows computations for a source of radius $r = 4\,mm$ and the detection plane at $z = 2.5\,mm$, and fig. 3.29 shows computations for a source of radius $r = 3\,mm$ and the detection plane at $z = 25\,mm$, showing agreement between the two computation methods.

As our second example we are going to study the elliptic disk as a non-rotational symmetric system. We stated that for a circular disk the radius is calculated with parameters $r = \left(\frac{m}{a}\right)^{1/2}$ and that the parameters m and p were equal. We fixed the parameter $a = 1$, and therefore the radius $m = r^2$ and $m = p$. Using the same set of parameters but making $m \neq p$ it is possible to analyze the edge ray cones of an elliptic disk source located at $z = 0$. Then for this case the parameters of the matrix 3.94 are $a = 1, b_1 = b_2 = b_3 = k = $

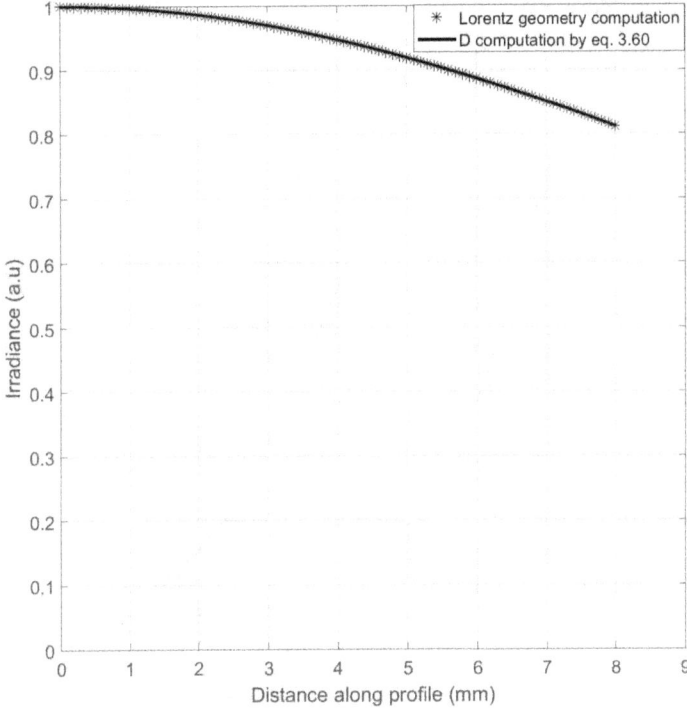

FIGURE 3.29: Comparison of Irradiance obtained at $z = 25\,mm$ for a circular disk of radius $r = 3\,mm$ between the Lorentz geometry method and eq. 3.75

$l = n = 0$ and $m = \alpha^2$ and $p = \beta^2$, where α and β are the semiaxes of the elliptic disk. With these parameters G^{-1} takes the form

$$G^{-1} = \begin{pmatrix} -x^2 + m & -xy & -xz \\ -xy & -y^2 + p & -yz \\ -xz & -yz & -z^2 \end{pmatrix}. \qquad (3.102)$$

The procedure to obtain the irradiance at any plane $z =$ constant is the same as for a circular disk: first we can obtain the direction of $\vec{D}$ by computing the eigenvector associated with negative eigenvalue of the matrix 3.102, and the modulus of $\vec{D}$ can be obtained by computing the two positive eigenvalues. Remember that the two positive eigenvalues are the semiaxes of the ellipse which cuts the cone of edge rays at a distance 1 from the computing point. Then the two angles of the elliptic cone of edge rays and finally the modulus can be obtained by the expression 3.9. In this case we have performed a comparison between results obtained by the application of the Lorentz geometry technique and raytracing computation. Figure 3.30 shows the comparison

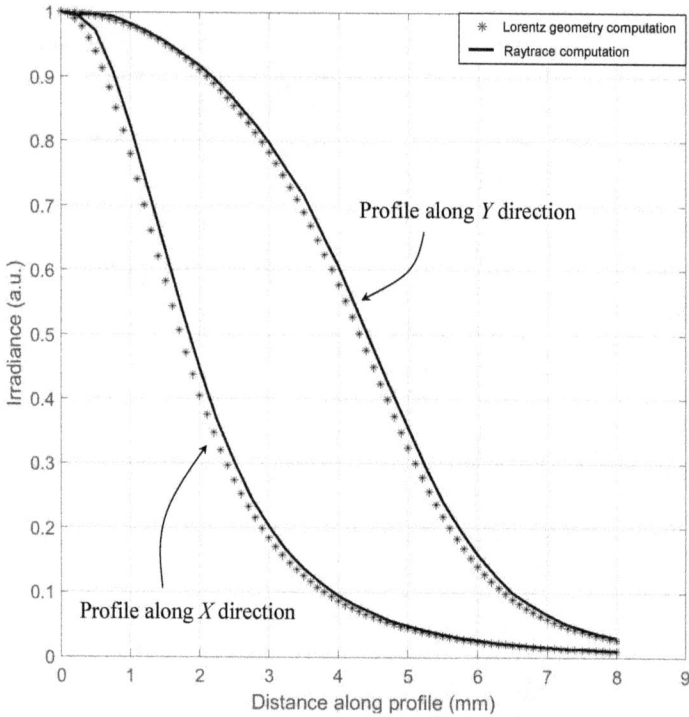

FIGURE 3.30: Comparison of vertical and horizontal irradiance profiles at plane $z = 3\,mm$ between the Lorentz geometry method and raytracing simulations, for a elliptic disk of semiaxes $\alpha = 1\,mm$ and $\beta = 5\,mm$

between the two techniques for a source with $\alpha = 1\ mm$, $\beta = 5\ mm$ and at a detection plane $z = 3\ mm$; the figure shows vertical and longitudinal irradiance profiles at the detection plane, and also shows agreement between the Lorentz geometry technique and raytracing simulations.

4

The Irradiance Vector in Optical Media

Refractive nonimaging optical devices have been developed starting many years ago. Refractive CPC using total internal reflection was described by Winston [47]: the maximum theoretical concentration ratio is increased by a factor n^2 using this device . On the other hand, design techniques for refractive nonimaging optics design have been also developed: simultaneous multiple surface (SMS) technique was developed by Benitez et al. [48], by which it is possible to design ideal refractive devices which match a source with a detector. Nevertheless, only a few concepts about the $\vec{D}$ vector in refractive media have been studied [49][9]. As an introductory notion, in a 2D system it is possible to analyze $\vec{D}$ for systems with refractive components using the basic property (section 1.4) that the difference in optical path length between edge rays is constant along a field line of $\vec{D}$. This remains true even in the presence of refractive media, fig. 4.1, where $[AP] - [A'P] = const$ and brackets indicate optical path lengths. It is possible to construct concentrators by placing mirrors along the field lines. Nevertheless this basic technique only provides information about the direction of $\vec{D}$. In this chapter we are going to study the properties of $\vec{D}$ in refractive media, including some techniques to compute $\vec{D}$ for optical systems with refractive or reflective optical components .

4.1 $\vec{D}$ Vector at Interface between Refractive Media

Let us consider a smooth, perfect transmission interface between two optical media with different refractive indices. No absorption or reflection, only refraction will be analyzed. Take a solenoidal $\vec{D}$ field vector incident on that surface: we can define the incident plane at point P as the plane formed by $\vec{D}$ and $\vec{N}$, where $\vec{N}$ is the vector normal to the interface at point P, fig. 4.2. By the symmetry of the problem, the refracted vector $\vec{D}'$ lies in the incident plane and we can write the vector in the form

$$\vec{D} = \vec{D}_n + \vec{D}_\tau, \tag{4.1}$$

DOI: 10.1201/9780367551605-4

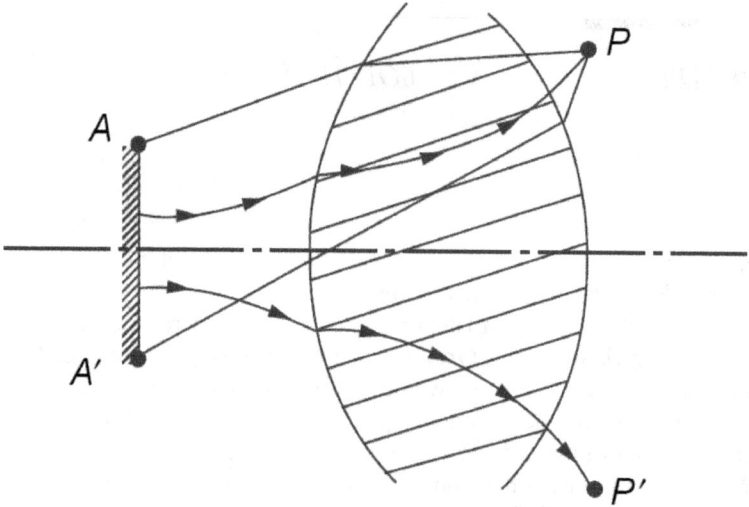

FIGURE 4.1: Field lines of $\vec{D}$ with refractive media. AA' is a Lambertian source. The arrows represent field lines, the plain lines rays

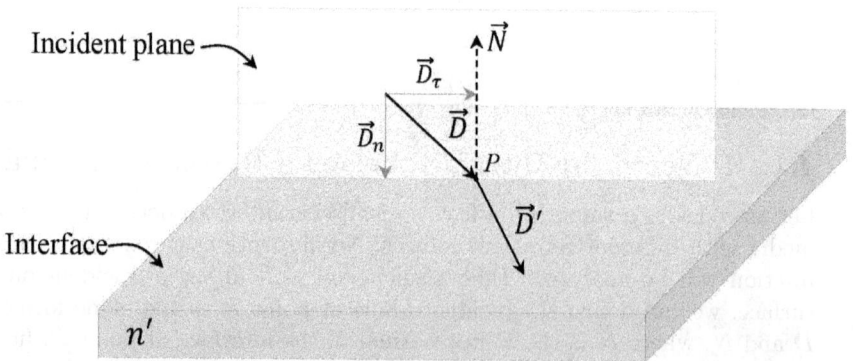

FIGURE 4.2: Decomposition of $\vec{D}$ at the interface between air and optical media with the refraction index n'

where $\vec{D}_n$ is the normal component parallel to $\vec{N}$, and $\vec{D}_\tau$ is the tangential component , orthogonal to $\vec{N}$ in the incident plane. It is possible to define a small cylindrical volume dv around the incident point P to which we can apply eq. 1.20. As the height of the cylinder approaches zero, we can write

$$\vec{D} \cdot d\vec{A} = \vec{D}' \cdot d\vec{A}, \tag{4.2}$$

where $d\vec{A}$ is the differential surface vector at the incident point P. Equation 4.2 shows that the flux is constant at both sides of the interface as a consequence of energy conservation. From this equation we can conclude that the normal component of $\vec{D}_n$ is continuous across the interface

$$\vec{D}_n = \vec{D}'_n. \tag{4.3}$$

Regarding the tangential component, in general it is not continuous at the interface $\vec{D}_\tau \neq \vec{D}'_\tau$ and must be computed in each situation. A particular case appears for interface surfaces orthogonal to $\vec{D}$, or what is the same, interfaces with the geometry of quasipotential surfaces, as we defined in section 3.7. For these special cases the vector $\vec{D}$ has only the normal component and is continuous across the interface. We will study these special cases in next section.

On the other hand, it is possible to study $\vec{D}$ at the interface in the case of 2D optical systems. For these systems it is easy to obtain the refracted vector $\vec{D}'$ remember that for 2D systems the direction of $\vec{D}$ points along the bisector of edge rays and the modulus of $\vec{D}$ is proportional to $\sin \theta_{cone}$, where θ_{cone} is the semiangle of the cone produced by edge rays at refraction point P; see fig. 4.3. For this configuration it is possible to apply Snell's law to two edge rays, r_1 and r_2:

$$n \sin(\theta_1) = n' \sin(\theta'_1) \tag{4.4}$$

$$n \sin(\theta_2) = n' \sin(\theta'_2). \tag{4.5}$$

By subtracting and adding both equations and applying trigonometric relations for the subtraction and addition of trigonometric sin functions, we have the pair of equations

$$n \sin\left(\frac{\theta_1 + \theta_2}{2}\right) \cos\left(\frac{\theta_1 - \theta_2}{2}\right) = n' \sin\left(\frac{\theta'_1 + \theta'_2}{2}\right) \cos\left(\frac{\theta'_1 - \theta'_2}{2}\right) \tag{4.6}$$

$$n \sin\left(\frac{\theta_1 - \theta_2}{2}\right) \cos\left(\frac{\theta_1 + \theta_2}{2}\right) = n' \sin\left(\frac{\theta'_1 - \theta'_2}{2}\right) \cos\left(\frac{\theta'_1 + \theta'_2}{2}\right). \tag{4.7}$$

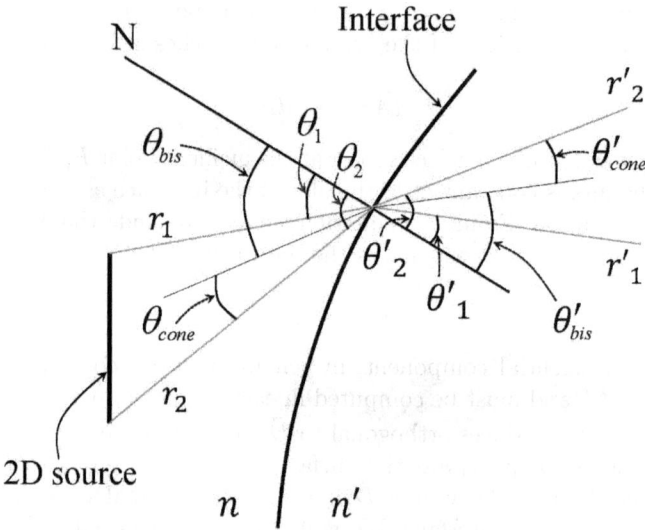

FIGURE 4.3: Refraction of $\vec{D}$ at the interface of 2D optical element

Then considering that $\theta_{bis} = \frac{\theta_1 + \theta_2}{2}$ points in the bisector direction and $\theta_{cone} = \frac{\theta_1 - \theta_2}{2}$ provides the semiangle of the cone, we can write

$$n \sin\left(\theta_{bis}\right) \cos\left(\theta_{cone}\right) = n' \sin\left(\theta'_{bis}\right) \cos\left(\theta'_{cone}\right) \tag{4.8}$$

$$n \cos\left(\theta_{bis}\right) \sin\left(\theta_{cone}\right) = n' \cos\left(\theta'_{bis}\right) \sin\left(\theta'_{cone}\right). \tag{4.9}$$

Solving this pair of equations for the two angles θ'_{bis} and θ'_{cone}, which are the unknown angles, it is possible to obtain the direction and the modulus of the refracted vector $\vec{D}'$ at any point on the interface for 2D optical systems.

4.2   Orthogonal Refractive Interfaces

In this section we are going to study particular interface geometries, those which are orthogonal to $\vec{D}$ at any point. In section 3.7, we called these surfaces quasipotential surfaces. We will use the concept of the flux tube in our study, defined as a tube such that the field vector $\vec{D}$ is tangent to the walls of the tube at any point. The main property of a flux tube is that the light flux through a flux tube containing no sources or sinks is constant, and this property is the

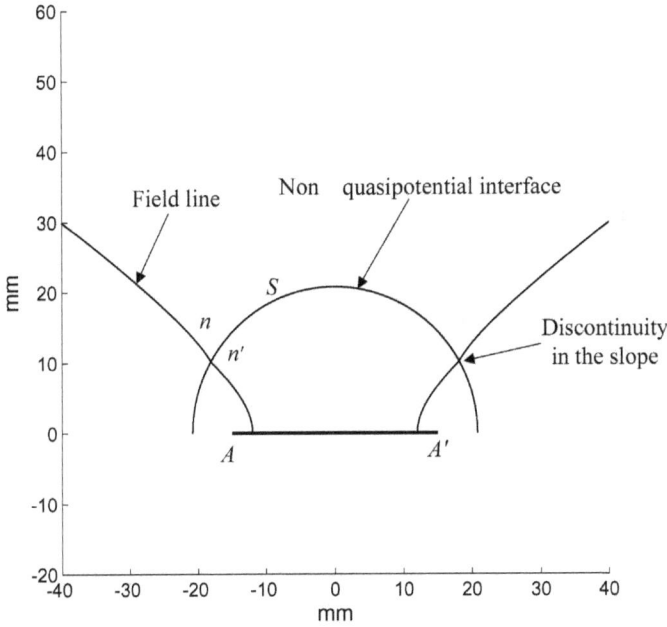

FIGURE 4.4: Field line of $\vec{D}$ produced by a Lambertian disk and a spherical diopter, $n' = 1.51$ and $n = 1$, non-quasipotential interface

principle behind using the flowline method to design concentrators. We can try to build a flux tube for the system of a Lambertian disk source and a smooth interface S between two transparent, homogeneous media with refractive index n and n'. If the interface between refractive media is not a quasipotential surface, for points at S the slope of $\vec{D}$ has a discontinuity and it is not possible to define the tangent of $\vec{D}$ vector fig. 4.4. Therefore it is not possible to build a flux tube with the walls tangent to $\vec{D}$ for all points, and in general the flux through tubes for these non-quasipotential optical systems is not conserved. Nevertheless, if the interface S is a quasipotential surface, at the interface $\vec{D}$ has only the normal component $\vec{D} = \vec{D}_n$. The slope of $\vec{D}$ is continuous through the interface and it is possible to define flux tubes with walls tangent to $\vec{D}$ at all points, which conserves the total flux, fig. 4.5. Then a method to build ideal dielectric concentrators is to produce a flux tube that generates quasipotential lines along the interface between the refractive media.s

To prove this result, as a first example we will study refractive concentrators produced by a Lambertian disk [49]. It is well known that field lines produced by a Lambertian disk are obtained by rotational symmetry of confocal hyperbolas, which are one-sheeted hyperboloids. The surfaces orthogonal to them are surfaces obtained by rotating confocal ellipses about its minor

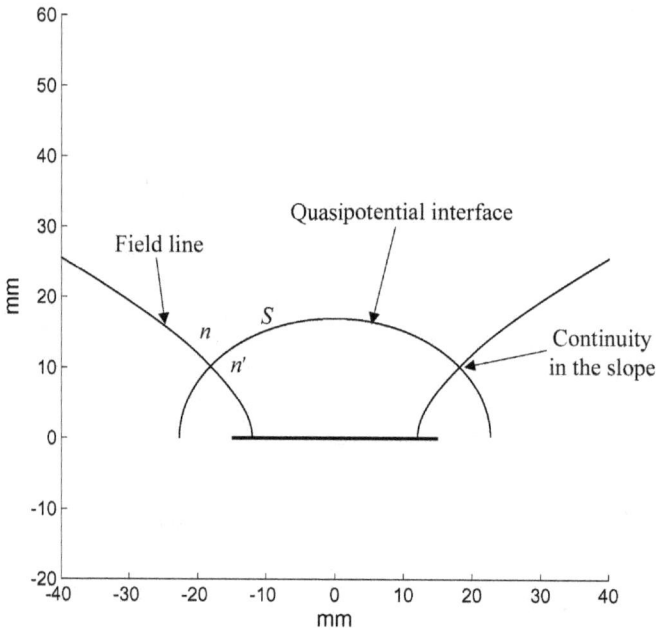

FIGURE 4.5: Field line of $\vec{D}$ produced by a Lambertian disk and an elliptical diopter, $n' = 1.51$ and $n = 1$, quasipotential interface

semiaxis, which are oblate spheroids. Then it is possible to build concentrators as flux tubes with perfect reflective walls with the shape of field lines and dielectric interfaces in the tube. Field lines inside dielectric material n' are hyperbolas with foci at A and A'. Outside of dielectric material, considering that $n = 1$, field lines can be calculated using the property that $\vec{D}$ points in the direction of the bisector between edge rays, and edge rays can be calculated using Fermat's principle. The concentrator can be obtained by rotational symmetry of the profile, fig. 4.6.

A quasipotential concentrator has a continuous slope at the reflective wall, whereas for a non-quasipotential concentrator there is a discontinuity in the slope of the field line at the intersection point between the refractive surface and field line. Despite this point of discontinuity, all other points of the non-quasipotential concentrator walls have the shape of field lines, and it works ideally. From a practical point of view it is expected that finite source non-quasipotential concentrators work close to ideally. We have analysed this situation by raytracing simulations: first, for a non-quasipotential concentrator produced by rotating symmetrically the profile of fig. 4.4, calculated using a Lambertian disk of radius $R_L = 15\,mm$ with a spherical diopter with refractive index $n' = 1.51$ and radius $R_d = 20.81\,mm$. The reflective flux tube

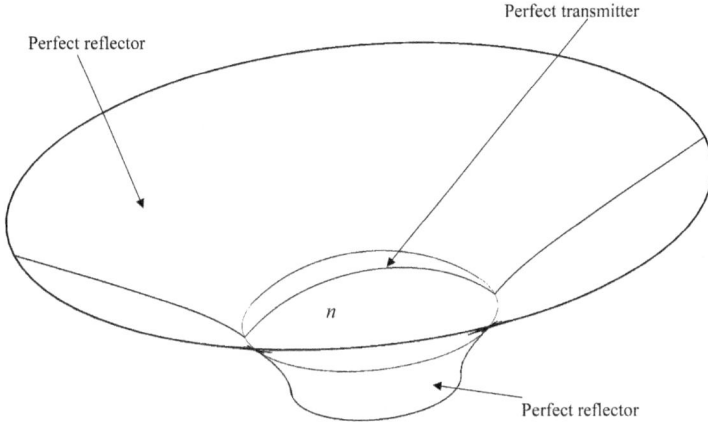

FIGURE 4.6: Quasipotential refractive concentrator built by rotational symmetry of profile in fig. 4.5; interface is an oblate spheroid.

used as concentrator has an entry aperture of radius $R_{in} = 40\,mm$ and exit aperture of radius $R_{out} = 12\,mm$. And second, for a quasipotential concentrator produced by rotational symmetry of profile of fig. 4.5, calculated using as a source a Lambertian disk of radius $R_L = 15\,mm$ and an oblate spheroid diopter of major semiaxis $a = 22.67\,mm$, minor semiaxis $b = 17\,mm$ and refractive index $n = 1.51$. For this configuration the étendue conservation provides a theoretical maximum transmittance

$$T_{max} = \left(\frac{nR_{out}}{R_{in}}\right)^2 = 0.2052, \tag{4.10}$$

obtaining the transmittance for a non-quasipotential concentrator, $T = 0.197$, by raytracing simulation, while for a quasipotential concentrator the maximum theoretical transmittance, eq. 4.10, was obtained with a small error of $7 \cdot 10^{-4}$. Applications for secondary concentrators in solar energy devices can be considered for this refractive concentrator.

As a second example, we have considered 2D CPC as an ideal infinite source concentrator. Figure 4.7 shows field lines and quasipotential lines produced by a virtual source in the shape of a truncated wedge which has two wings infinitely extended at an angle of $2\theta = 60°$. The field lines which start at the vertex of the wedge produce 2D CPC, and the other field lines produce HPC. Quasipotential lines are orthogonal to the field lines; in region I quasipotential lines are horizontal lines; in region II they are tilted parabolas with foci at A'; and in region III they are ellipses with A and A' as foci. Therefore

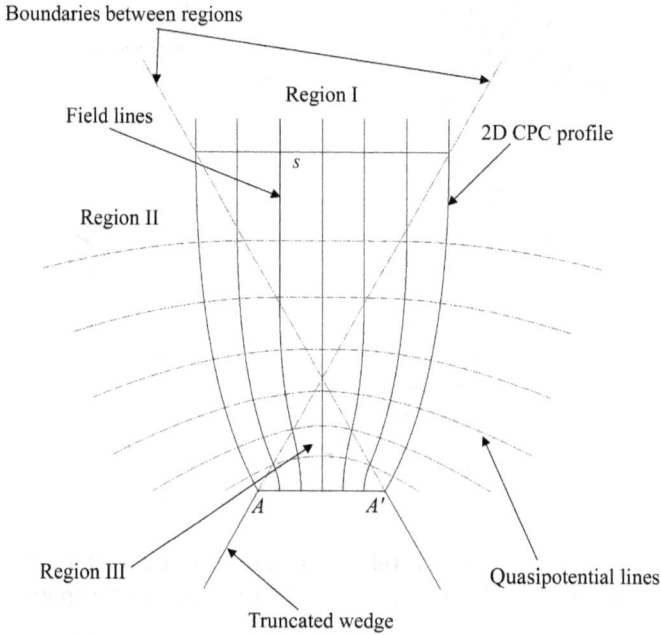

FIGURE 4.7: Field lines and quasipotential lines produced by
a truncated wedge.

the way to build an ideal refractive concentrator is to guarantee that the interface between media has the shape of a quasipotential surface. In this case,
because 2D CPC fits this description, we can use its field lines to build the
concentrator. The quasipotential line, S in fig. 4.7, is a flat line in region I.
It is well known that 2D refractive CPC with a flat surface, or quasipotential
surface, as the interface of refractive media achieves the maximum concentration ratio [50]. But it is interesting to consider other interface geometries, to
compare between quasipotential surfaces and non-quasipotential surfaces for
a 2D refractive CPC, and to prove the theoretical result. We have simulated
the transmission angle curves for several 2D refractive CPC, all of them with
perfect mirrors as walls, designed with wedge semiangle $\theta_s = 30°$, exit aperture width 15 mm, refractive index $n' = 1.51$ and different extruded refractive
interface geometries replacing flat surfaces. Figure 4.8 shows the results obtained by raytracing simulations . The dashed dot line shows the transmission
angle curve for a curved refractive interface of radius 200 mm; the same transmission angle data were obtained for convex and concave curved interface with
the same radius. The dashed line shows the transmission angle for a curved
refractive interface of radius 400 mm, and the continuous line shows the simulated transmission angle for a quasipotential flat interface, which achieves
ideal behaviour with a cutoff angle $\theta_c = \arcsin(n' \sin \theta_s) = 49°$.

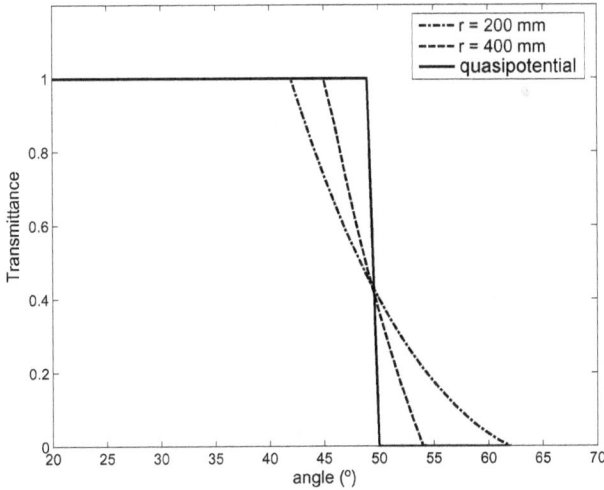

FIGURE 4.8: Transmission angle curves for 2D refractive CPC with different interface geometries

4.3 Refracted Cone of Edge Rays

In section 1.5 we introduced the concept of the cone of edge rays , and from the edge ray principle , we stated the important property that this cone of edge rays provides all information to obtain the modulus and direction of $\vec{D}$. Now we are going to generalize this concept of cone of edge rays and apply it to optical systems with refractive or reflective elements. Let us start with a Lambertian source without refractive or reflective elements, fig. 4.9.We want to compute the $\vec{D}$ at any point P. As we saw in section 3.3, the more efficient way to compute $\vec{D}$ is by using eq. 3.32,

$$\vec{D} = \frac{B}{2} \sum_{i=1}^{N} \alpha_i \vec{n}_i, \qquad (4.11)$$

with N being the number of segments into which we divide the contour of the source. Equation 4.11 needs to compute only the edge rays produced at the vertex of the segments into which we divide the contour of the source. For a circular source we can approach the circular contour by using a segmented contour with number of sides great enough to obtain the desired resolution, as we showed in section 3.4.

When a refractive medium is introduced, the effect of the medium is to transform the cone of edge rays by redirecting it and reshaping it in such a way that an observer at point P receives radiation only from what we have called

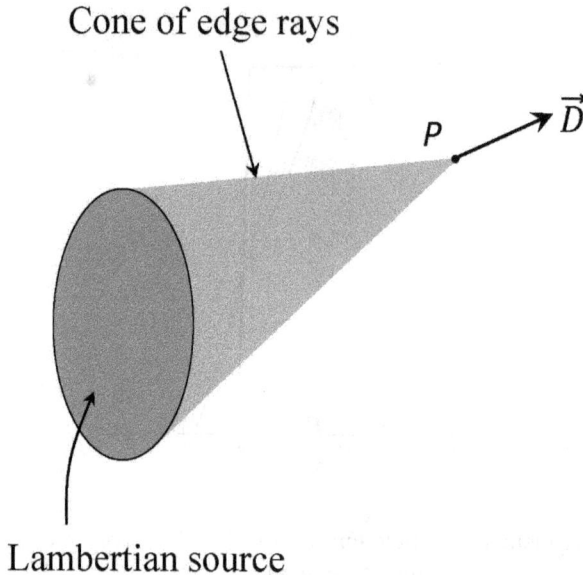

Cone of edge rays

P

$\vec{D}$

Lambertian source

FIGURE 4.9: Cone of edge rays for a simple Lambertian source

the refracted cone of edge rays, fig. 4.10. An analogous situation appears for a reflective surface , and in that case we call it the reflected cone of edge rays. The refracted/reflected cone of edge rays can be obtained by the application of Fermat's principle to edge rays, and then it is possible to compute $\vec{D}$ in refractive/reflective media applying eq. 4.11 to the refracted or reflected cone of edge rays. Using Fermat's principle it is possible to obtain the directions and the angles of refracted or reflected edge rays. Once we have the direction and the angles of the refracted/reflected cone of edge rays, by applying eq. 4.11 to the refracted/reflected cone we can obtain $\vec{D}$ at point P inside the refractive medium.

Nevertheless, to apply Lorentz geometry and matrix G in optical media it is necessary to define the concept of contour of refracted/reflected cone of edge rays, fig. 4.11. We can see this, for an observer at point P inside the refractive medium, as the contour from which the radiation appears to originate. The question is, where is this contour located? It is possible to answer this question by using energy conservation. To do this we can state that the energy of the radiation that arrives at P from the Lambertian source must be the same as the energy of the radiation that arrives at P from the contour of the refracted cone. Note that the contour of the refracted cone is in the refractive medium of index n, and the Lambertian source is in the air. The

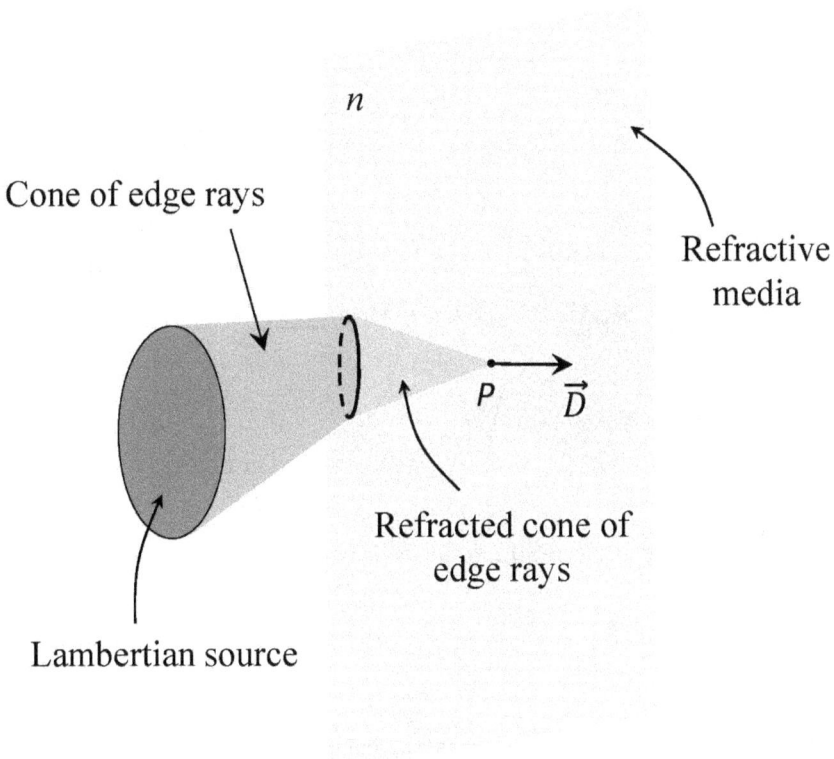

FIGURE 4.10: Refracted cone of edge rays to compute $\vec{D}$ at point P in refractive media

energy that travels from the interface to an infinitesimal area located at point P and orthogonal to $\vec{D}$ must be the same, by construction, from Lambertian and from the contour of the refracted cone. Therefore it is necessary to impose conservation of energy at the interface of the refractive media between the radiation from the Lambertian source and the radiation from the contour of the refracted cone of edge rays ; this provides us the location of the contour of the refracted cone.

To illustrate the concept of the contour of the refracted cone of edge rays, let us take a circular Lambertian source and a plane diopter. We can analyze the $\vec{D}$ at the axis. For points at the axis, by symmetry, the contour of the refracted cone of edge rays will be also circular and centered at the axis, fig. 4.12. Then at the axis all the cones of edge rays will be circular cones. For

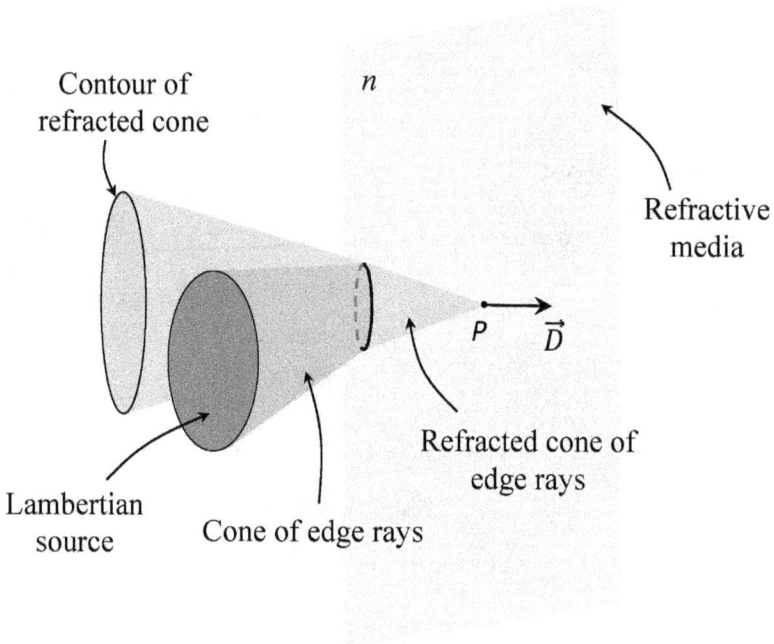

FIGURE 4.11: Contour of the refracted cone of edge rays to compute $\vec{D}$ at point P in refractive media

this simple configuration it is possible to obtain analytic results based on the contours of the refracted cones of edge rays.

Let us consider the 2D sketch of fig. 4.13, with the Z axis being the symmetry axis of the source. The source has a radius R, and its center is located at $z = 0$; the plane diopter is located at $z = d$ and has a refractive index n. Using Snell's law it is possible to deduce the expression of coordinate z_{apex} of the apex of the refracted cone of edge rays as a function of the incident coordinate x_i at the interface with the plane diopter.

$$z_{apex} = d + x \left(\frac{n^2 d^2}{(R-x)^2} + n^2 - 1 \right)^{1/2} \tag{4.12}$$

We can impose the conservation of energy, in the sense that the radiation incident at the circular interface in a plane diopter from the source must be the same as that radiation incident at the circular interface in a plane diopter from the contour of the refraction cone of edge rays. To do that we must take into account the concept of basic radiance, which comes from the conservation of étendue, [51][52], which states that the brightness of a radiation beam inside

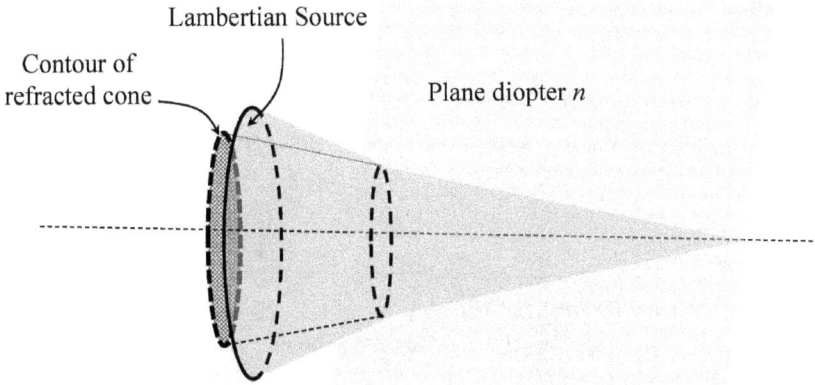

FIGURE 4.12: Contour of a refracted cone for a circular Lambertian source and a plane diopter, for point P at axis

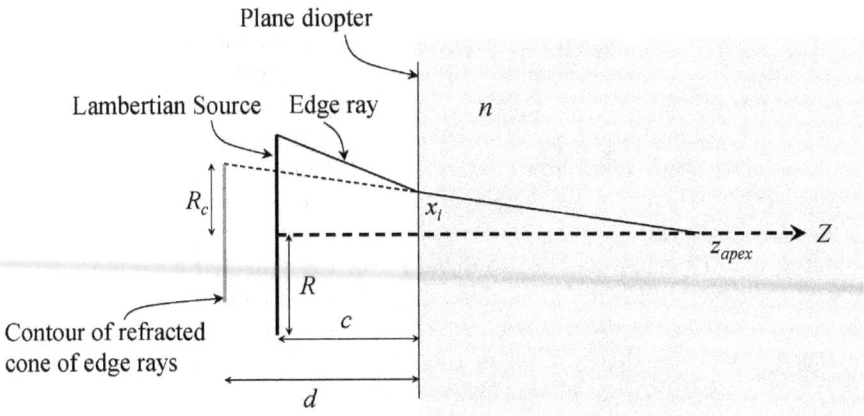

FIGURE 4.13: 2D sketch to evaluate the contour of a refracted cone in the circular source plane diopter system

refractive media n_1 is related to the brightness of the same radiation beam inside other refractive media n_2 by

$$\frac{B_1}{n_1^2} = \frac{B_2}{n_2^2}.$$ (4.13)

Now we can state the conservation of energy between the source and the contour of the refracted cone of edge rays, at the interface of a plane diopter, by

$$B_2 F_{ci} = B_1 F_{si}$$ (4.14)

where F_{ci} is the view factor from the contour of the refracted cone of edge rays to the interface and F_{si} is the view factor from the circular Lambertian source to the interface [53]. The view factor between two coaxial circular apertures is well known [54]; for example, consider fig. 4.13.

$$F_{si} = \frac{1}{2}\left(1 + \frac{c^2}{R^2} + \frac{x_i^2}{R^2} - \sqrt{1 + \frac{c^2}{R^2} + \frac{x_i^2}{R^2} - 4\frac{x_i^2}{R^2}}\right)$$ (4.15)

Then from eq. 4.14 it is possible to obtain F_{ci}, and considering the simple relation between the radius of the contour of the refracted cone R_c and the distance between the contour of the refracted cone and the interface d,

$$R_c = x_i + d\tan(\theta_c),$$ (4.16)

it is possible to obtain the parameters R_c and d of the contour of the refracted cone. We have checked this result with raytracing simulations, by comparison of the irradiance at the axis produced by the system with the source and plane diopter and the irradiance at the axis produced from the contour of the refracted cone, taking into account basic radiance eq. 4.13. Following the sketch in fig. 4.13 we use a source of radius $R = 5\,mm$; the plane diopter is placed at $c = 20\,mm$ with a refractive index $n = 1.5$, fig. 4.14.

Another trivial example of the contour of the reflected cone is that produced by a plane mirror. For a plane mirror we can compute $\vec{D}$ at any point in space as if there were a virtual source placed symmetrically to the mirror plane. We will use this concept of the contour of the refracted cone of edge rays in the final sections of this chapter, where we will study the application of Lorentz geometry to analyze systems with refractive or reflective elements.

4.4 $\vec{D}$ through Refractive Media

In this section we are going to present the method and examples to compute $\vec{D}$ for optical systems with refractive elements. The method is completely

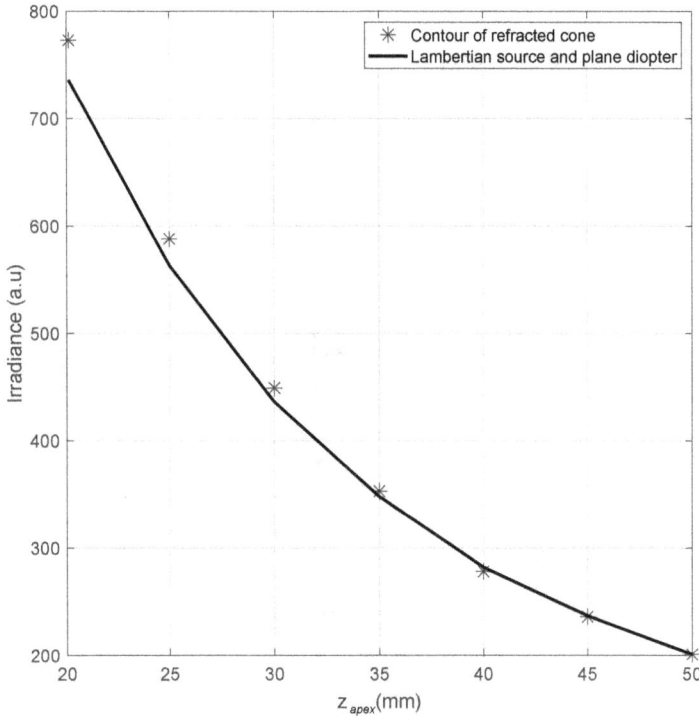

FIGURE 4.14: Raytrace simulation to check the concept of the contour of refracted cone of edge rays

general, regardless of whether there is an optical axis or not or the number of refractive elements. It basically consists of obtaining the structure of the refracted cone of edge rays using Fermat's principle ; once we have the refracted cone of edge rays we compute $\vec{D}$ by means of eq. 4.11; and finally we obtain the irradiance at a detector plane parallel to plane XY using the D_z component. This can be done using mathematical computation software, and we compare results from $\vec{D}$ computation with raytracing simulations. We have done this for a simple system of a square source and a plane convex lens with refraction index $n = 1.5$, the convex surface with radius of 300 mm, and we have analyzed the irradiance produced by this system at different detector positions, fig 4.15. The square source has a length of 40 mm and is located at $z = 0$, the lens is located at $z = 1255\,mm$ and the lens thickness is 30 mm. We have compared the computation of irradiance from $\vec{D}$ calculations and from raytracing simulations. Figure 4.16 shows normalized irradiance maps for a detector at $z = 1700\,mm$ from the source, fig. 4.16a) for $\vec{D}$ vector computation and fig. 4.16b) for raytracing simulation. Figure 4.17 shows the comparison of results for the central irradiance profile at the detector. Figure 4.18 shows

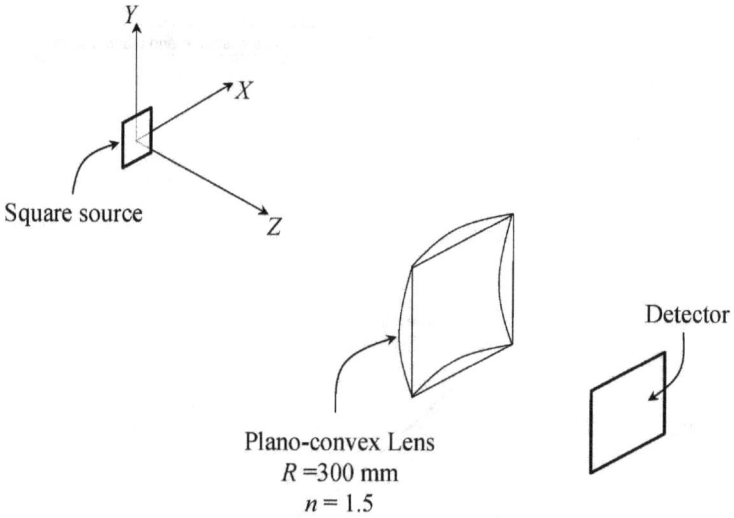

FIGURE 4.15: Arrangement to compute $\vec{D}$ through refractive media

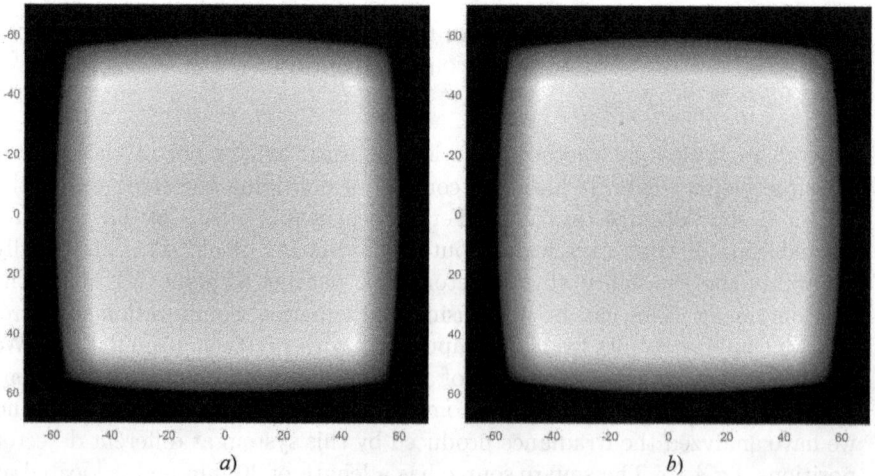

FIGURE 4.16: Irradiance map with the detector at $z = 1700\,mm$, a) $\vec{D}$ computation, b) raytracing simulation

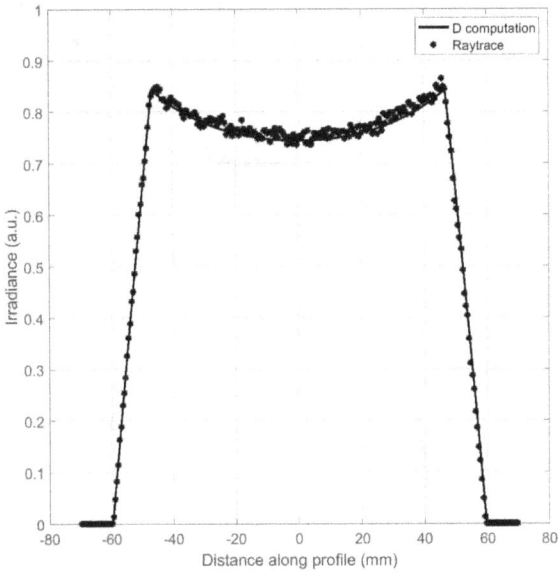

FIGURE 4.17: Irradiance profile comparison between $\vec{D}$ computation and raytracing simulation, with the detector at $z = 1700\,mm$

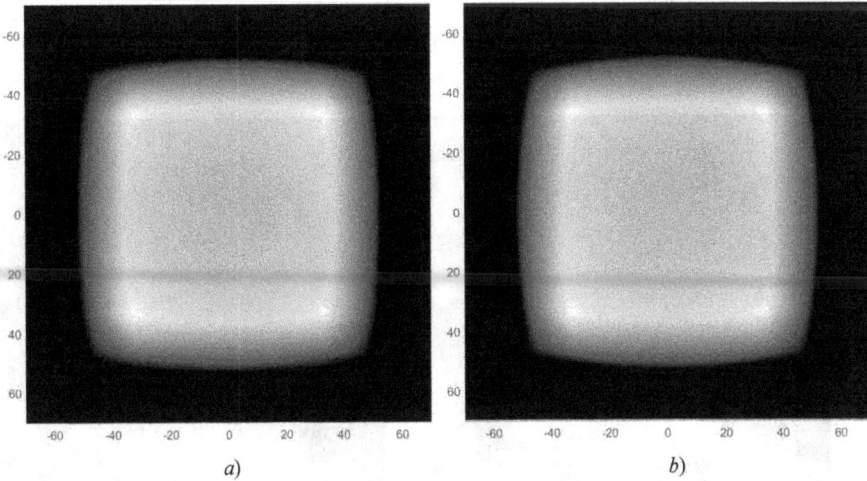

FIGURE 4.18: Irradiance map with the detector at $z = 1800\,mm$, a) $\vec{D}$ computation, b) raytracing simulation

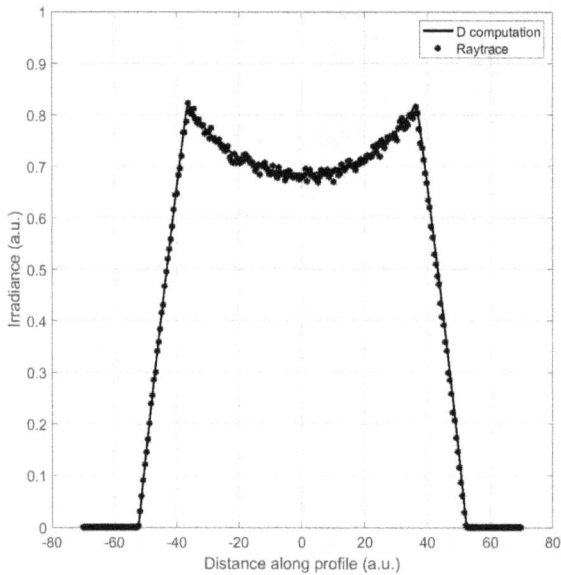

FIGURE 4.19: Irradiance profile comparison between $\vec{D}$ computation and raytracing simulation, with the detector at $z = 1800\,mm$

normalized irradiance maps for a detector at $z = 1800\,mm$ from the source, and Figure 4.19 shows the comparison of results for the central irradiance profile at the detector. Finally, fig. 4.20 shows normalized irradiance maps for a

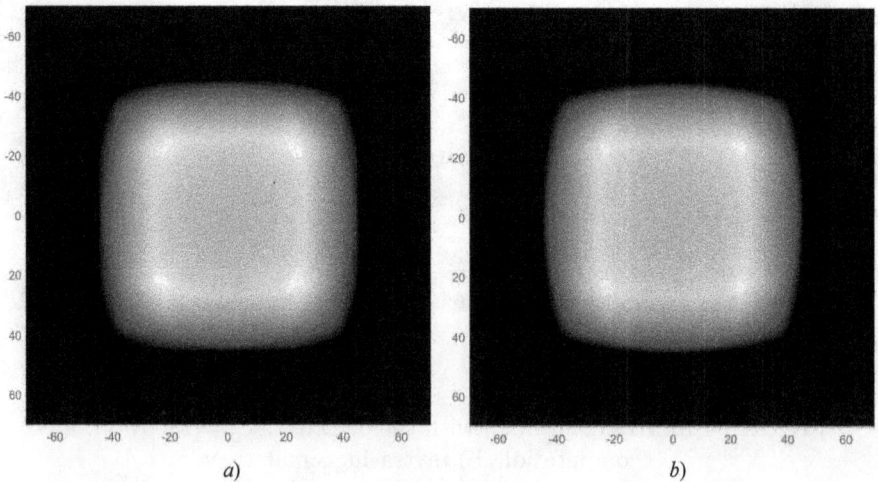

FIGURE 4.20: Irradiance map with the detector at $z = 1900\,mm$, a) $\vec{D}$ computation, b) raytracing simulation

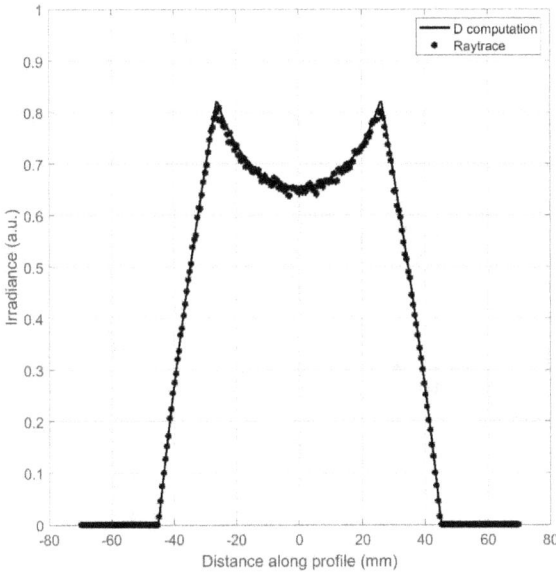

FIGURE 4.21: Irradiance profile comparison between $\vec{D}$ computation and raytracing simulation, with the detector at $z = 1900\,mm$

detector at $z = 1900\,mm$ from the source, and fig. 4.21 shows the comparison of results for the central irradiance profile at the detector.

4.5 $\vec{D}$ through Reflective Media

Now that we have analyzed the $\vec{D}$ in systems with refractive elements, we are going to analyze it for systems with reflective elements. The analysis is analogous to that of refractive media, and the results are also analogous. We need to evaluate the reflected cone of edge rays produced by the source also using Fermat's principle ; then, using eq. 4.11, we will compute the $\vec{D}$ at the detector. As usual we are going to use a detector in plane XY; then the D_z component of $\vec{D}$ is the irradiance at the detector. Figure 4.20 shows a sketch of the analyzed reflective system. We have studied a perfect spherical mirror with radius $R = 300$ mm, the center of the mirror located at $z = 110\,mm$ and with a square contour of $140\,mm$ per side. The source was a square Lambertian source of $10\,mm$ per side, located at the origin of the reference system and emitting radiation in the negative z direction. There was also a square detector of $80\,mm$ per side. We have analyzed the irradiance maps and irradiance profiles at the detector for two positions of the detector at $z = 260\,mm$ and at $z = 450\,mm$.

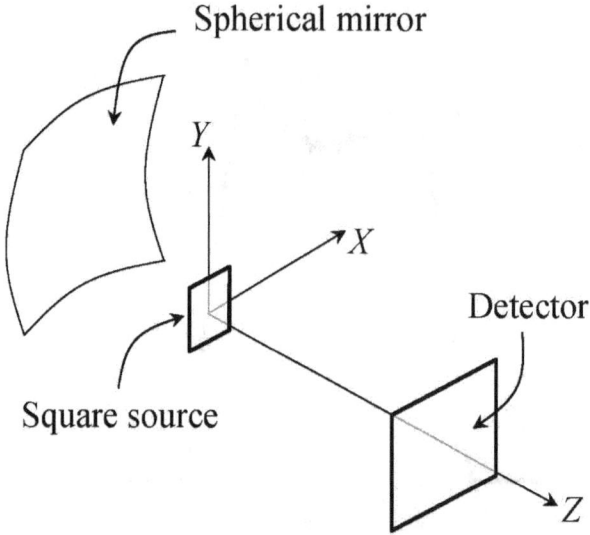

FIGURE 4.22: Arrangement to compute $\vec{D}$ through reflective media

Figure 4.23 shows normalized irradiance maps for a detector at $z = 260\,mm$ from the source; fig. 4.23a for $\vec{D}$ vector computation and fig. 4.23b for raytracing simulation. Figure 4.24 shows the comparison of results for

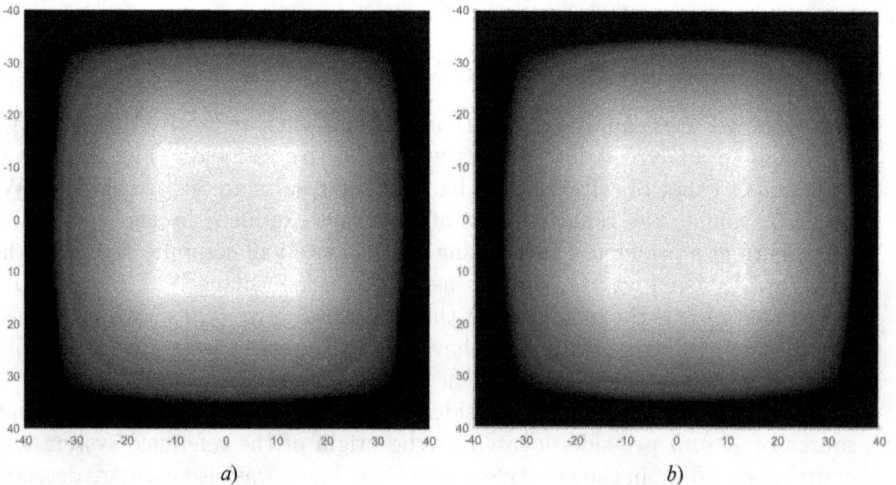

FIGURE 4.23: Irradiance maps for reflective optics system with the detector at $z = 260\,mm$: a) $\vec{D}$ computation, b) raytracing simulation

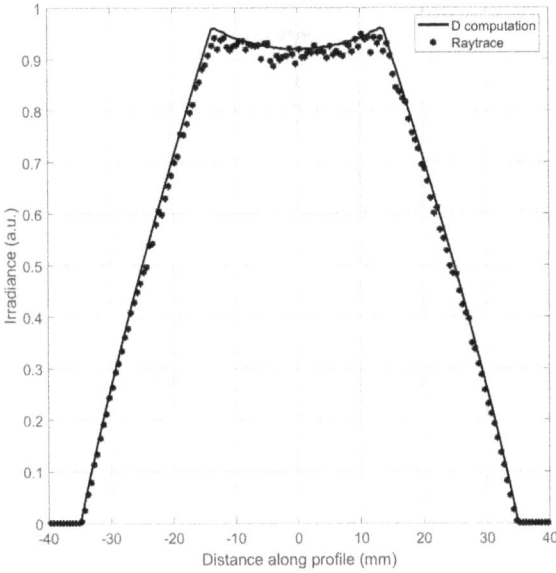

FIGURE 4.24: Irradiance profile comparison between $\vec{D}$ computation and raytracing simulation, with the detector at $z = 260\,mm$

the central irradiance profile at the detector, showing good agreement between $\vec{D}$ computation and raytracing simulation. Figure 4.25 shows normalized

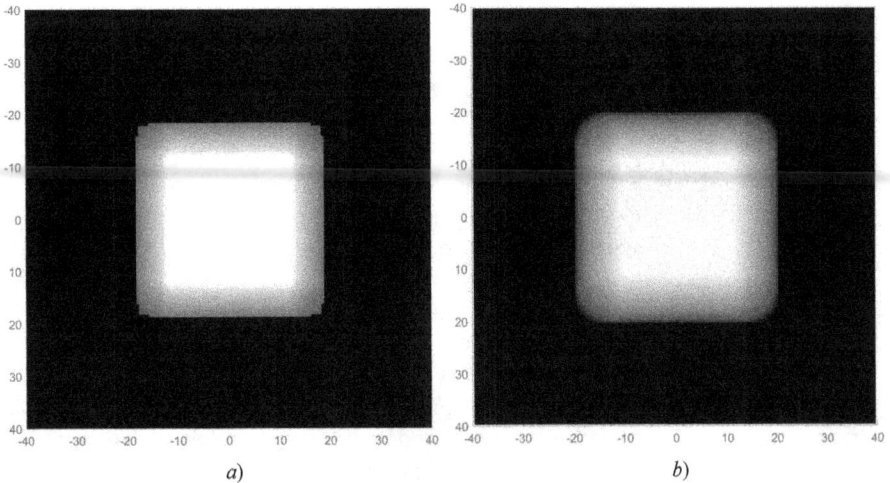

FIGURE 4.25: Irradiance maps for reflective optics system with the detector at $z = 450\,mm$: a) $\vec{D}$ computation, b) raytracing simulation

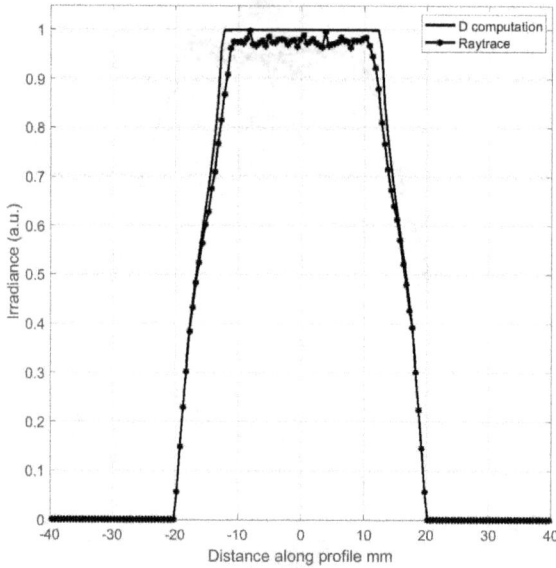

FIGURE 4.26: Irradiance profile comparison between $\vec{D}$ computation and raytracing simulation, with the detector at $z = 450\,mm$

irradiance maps for a detector at $z = 450\,mm$ from the source; fig. 4.25a for $\vec{D}$ vector computation and fig. 4.25b for raytracing simulation . Figure 4.26 shows the comparison of results for the central irradiance profile at the detector.

4.6   Lorentz Formalism to Compute $\vec{D}$ through Refractive Media

In section 3.9, we have applied the Lorentz technique , extensively used in general relativity theory, to the evaluation of $\vec{D}$ in free space propagation. In this and the following sections, we are going to apply Lorentz formalism to the analysis of optical systems with refractive and reflective media. It is interesting to remember that the rationality behind this is that Lorentz geometry provides a field of cones in space and provides the cone of edge rays at any point in the space P, which must be lightlike curves in the Lorentz metric.

Adopting a Lorentz geometry approach, let (M, g) be a Lorentz manifold, where M is an open subset in $\mathbb{R}^3$, and g be a symmetric twice covariant tensor field on M such that at each point $p \in M$, the bilinear form g_p : $T_p M \times T_p M \rightarrow \mathbb{R}$ is non-degenerate and has the signature $(+, +, -)$. As usual, a vector $v \in T_p M$ is called timelike if $g(v, v) < 0$, lightlike if $g(v, v) = 0$ and $v \neq 0$, and spacelike if $g(v, v) > 0$ or $v = 0$, and it is known as the causal character of the vector v. If $\gamma : \mathbb{R} \to M$ is a C^1 curve, then it is called timelike, lightlike or spacelike if all its tangent vectors have the corresponding causal character. The set of lightlike vectors in T_pM forms the cone of edge rays at P.

Let (x, y, z) be the standard coordinate system in $\mathbb{R}^3$, and $G(P)$ the matrix which consists of g_p components. As we saw in section 3.9, the matrix G can be diagonalized $(\lambda_1, \lambda_2, \lambda_3)$ where one of the eigenvalues is negative, say λ_3. Let $(\vec{u}_W, \vec{u}_V, \vec{u}_D)$ be a orthogonal basis B' of eigenvectors of G, with $\vec{u}_D$ being the eigenvector corresponding to negative eigenvalue λ_3. Let $\vec{v} \in T_pM$ be a lightlike vector, $\vec{v} = y_1\vec{u}_W + y_2\vec{u}_V + y_3\vec{u}_D$. The condition $g(\vec{v}, \vec{v}) = 0$ can be expressed in the basis B' by the equation

$$\lambda_1 y_1^2 + \lambda_2 y_2^2 + \lambda_3 y_3^2 = 0, \qquad (4.17)$$

which is a cone whose axis is parallel to the unit vector $\vec{u}_D$ (remember that $\lambda_3 < 0$ and $\lambda_1, \lambda_2 > 0$). The intersection of the cone and the plane $y_3 = 1$ is the ellipse

$$\lambda_1 y_1^2 + \lambda_2 y_2^2 = -\lambda_3. \qquad (4.18)$$

The ellipse axes are parallel to the unit vectors $\vec{u}_W$ and $\vec{u}_V$ (see fig. 3.27), and their respective semiaxis lengths are $(-\lambda_3/\lambda_1)^{1/2}$ and $(-\lambda_3/\lambda_2)^{1/2}$. As in section 3.6, we summarize here:

1. The cone of edge rays at point P is elliptic, $\forall P \in \Re$.

2. The direction of the elliptic cone axis is the direction of $\vec{D}$, with $\vec{u}_D$ (the unit vector in the direction of $\vec{D}$) being the eigenvector of G associated with the negative eigenvalue of G.

3. The directions of the principal axes of the elliptic cone are the directions of $\vec{u}_W$ and $\vec{u}_V$, the eigenvectors of G associated with the positive eigenvalues of G.

In section 3.9 we studied the case of a homogeneous medium with refractive index $n = 1$. In this section we are going to generalize that result to optical systems with refractive and reflective elements.

As a first example we are going to analyze a refractive optical system with rotational symmetry. Let us take the optical system of fig. 4.27, composed of a circular Lambertian disk, a circular plane-convex lens with a refractive index $n = 1.5$ and a detector. We will study this system in $z = $ constant planes located between the lens and the image plane, in order to have irradiance patterns with maximum and minimum values. As in sections 4.4 and 4.5 we will focus our attention on the concept of the refracted cone of edge-rays , fig. 4.10. Let there be a point P at the detector; radiation from a refracted cone of edge rays will arrive at P. Now it is useful to use the concept of the contour of the refracted cone of edge rays, introduced in section 4.3, as the contour from which the radiation appears to originate, fig. 4.11. The

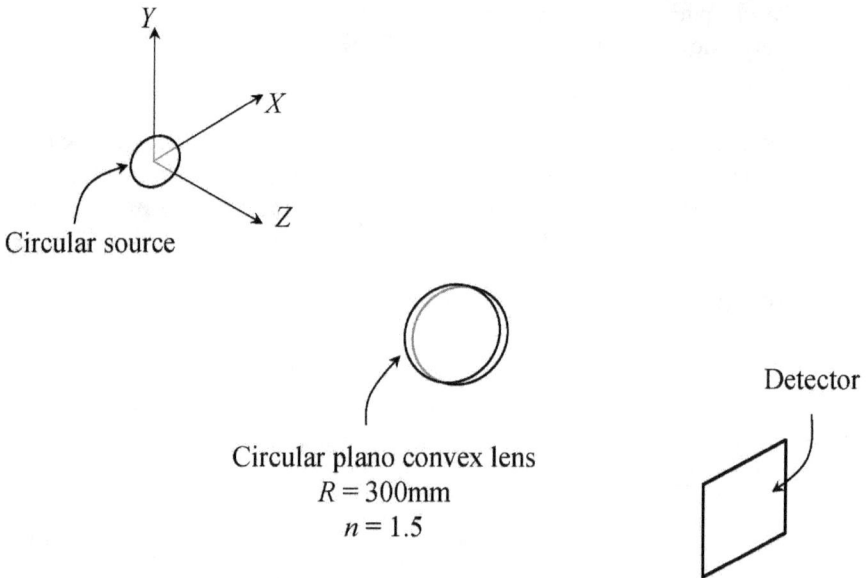

FIGURE 4.27: Sketch of the rotational symmetric refractive optical system analyzed by the Lorentz technique

contour of the refracted cone is different at any point on the detector, and its center is located at a point in the space (x_c, y_c, z_c), also different at any point on the detector. This contour of the refracted cone must be calculated for any point on the detector at which we want to evaluate $\vec{D}$. For a rotational symmetric optical system we can state that this contour of the refracted cone has an elliptic shape, and using Fermat's principle it is possible to evaluate the direction of the refracted cone of edge rays. Then by imposing the law of the conservation of energy, as we did in section 4.3, it is possible to evaluate the location of the contour of refracted cone of edge rays, for any point P at the detector, and then the semiaxes α and β of this elliptic shape. Once we have the contour of the refracted cone of edge rays it is possible to obtain the matrix G, eq. 3.102, one matrix G for each one of the points P at the detector. Then using the procedure of section 3.6 it is possible to evaluate the direction of $\vec{D}$ and its modulus with the eigenvalues $(\lambda_1, \lambda_2, \lambda_3)$. As in eq. 3.102 the parameters of the the matrix are $a = 1, b_1 = b_2 = b_3 = k = l = n = 0$ and $m = \alpha^2$ and $p = \beta^2$, where α and β are the semiaxes of the contour of the refracted cone of edge rays.

As an example, we can show the irradiance pattern for a system with a circular Lambertian source of radius $r_s = 20\,mm$, located at the origin, with axis Z being the symmetry axis; and a plano-convex lens with radius

FIGURE 4.28: Irradiance pattern obtained by raytracing of the system of
fig. 4.27 with the detector at $z = 1800$

$R_l = 300\,mm$, width of $30\,mm$, with the convex surface oriented to the source
and the vertex of the convex surface located at $z = 1255\,mm$. Finally, we have
analyzed this system for different positions of the detector plane. As usual, we
have compared the irradiance pattern obtained by raytracing with, in this case,
the irradiance pattern obtained by the Lorentz geometry technique explained
in this section. Figure 4.28 shows the irradiance pattern produced by this
system at $z = 1800\,mm$.

We have compared the irradiance profiles obtained by raytracing with
those obtained by the Lorentz geometry technique. Figure 4.29 shows the irradiance profiles for a detector at $z = 1800\,mm$, fig. 4.30 shows the irradiance
profiles for a detector at $z = 1900\,mm$ and fig. 4.31 shows the irradiance
profiles for a detector at $z = 2000\,mm$: the three figures show good agreement between raytracing irradiance profiles and Lorentz geometry irradiance
profiles.

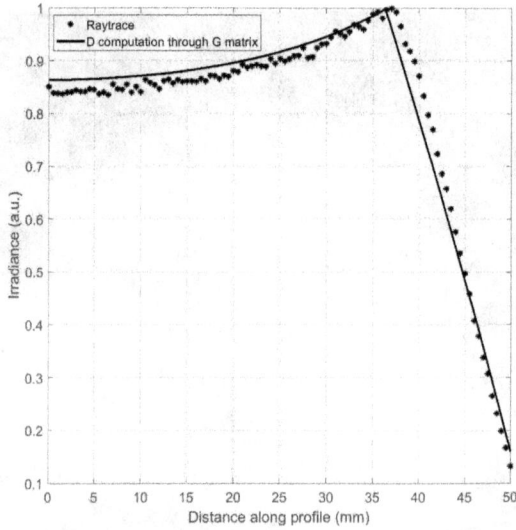

FIGURE 4.29: Comparison of irradiance profiles obtained by raytracing and by Lorentz geometry for the refractive system of fig. 4.27 with the detector at $z = 1800 \ mm$

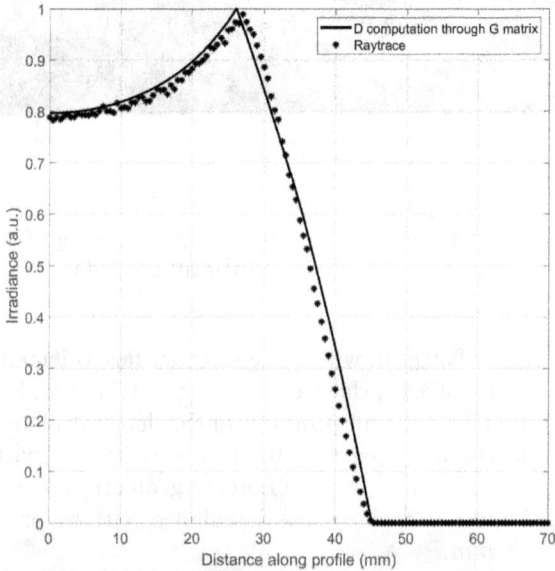

FIGURE 4.30: Comparison of irradiance profiles obtained by raytracing and by Lorentz geometry for the refractive system of fig. 4.27 with the detector at $z = 1900 \ mm$

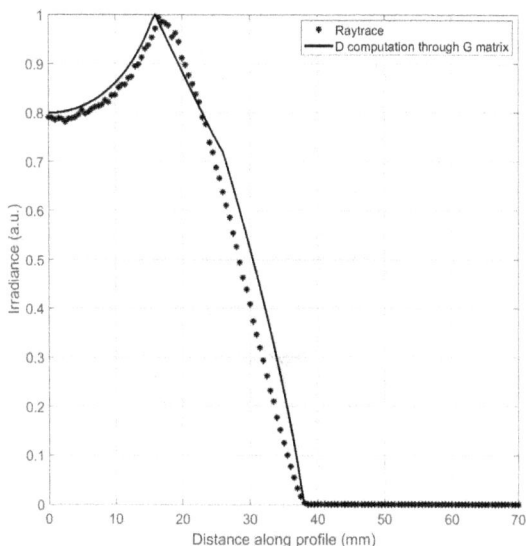

FIGURE 4.31: Comparison of irradiance profiles obtained by raytracing and by Lorentz geometry for the refractive system of fig. 4.27 with the detector at $z = 2000\ mm$

4.7 Using Lorentz Formalism to Compute $\vec{D}$ through Reflective Media

As second example we are going to analyze a reflective optical system with rotational symmetry. Let us take the optical system of fig. 4.32, composed of a circular Lambertian disk, a spherical mirror and a detector. The circular disk emits radiation in the negative direction of the Z axis; it is reflected by the spherical mirror and then detected at the detector. The procedure to obtain the irradiance distribution is the same as in the previous example: using Fermat's principle , we can evaluate the reflected cone of edge rays and the contour of the reflected cone of edge rays at any point on the detector P by imposing the law of conservation of energy. Because of the symmetry we can state that the contour of the reflected cone has an elliptic shape with semiaxes α and β, and its location can be obtained by imposing the conservation of energy between the contour of the reflected cone and the Lambertian source. Once we have the contour of the reflected cone, it is possible to evaluate G, one matrix G for each one of the points P at the detector. Then using the procedure of section 3.9 it is possible to evaluate the direction of $\vec{D}$ and its modulus with the eigenvalues $(\lambda_1, \lambda_2, \lambda_3)$. As in eq. 3.102 the parameters of the the matrix are $a = 1, b_1 = b_2 = b_3 = k = l = n = 0$ and $m = \alpha^2$ and

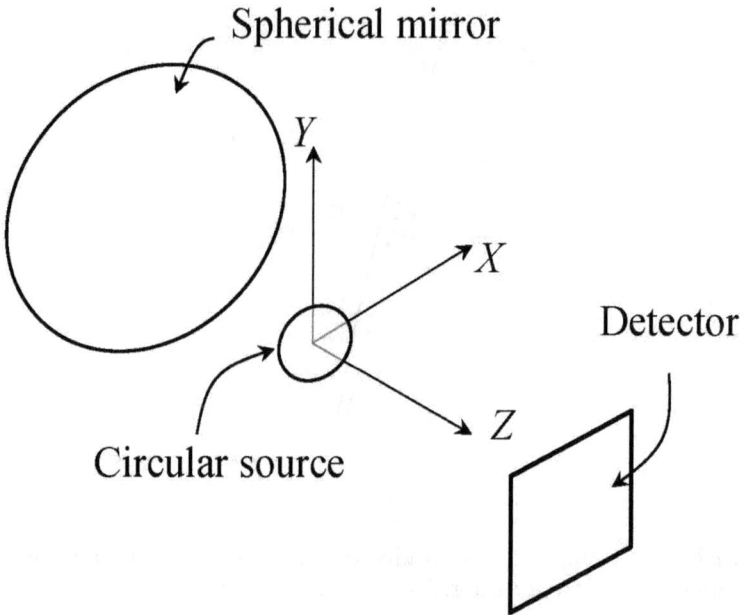

FIGURE 4.32: Sketch of the rotational symmetric reflective optical system analyzed by the Lorentz technique

$p = \beta^2$, where α and β are the semiaxes of the contour of the reflected cone of edge rays.

As an example, we can show the irradiance pattern of a system with a circular Lambertian source of radius $r_s = 10\,mm$, located at the origin, with axis Z being the symmetry axis; and a spherical mirror with radius $R_m = 400\,mm$, the center of the mirror located at $z = 140\,mm$ and the vertex located at $z = -260\,mm$. We have analyzed this system for a detector at $z = 450\,mm$. As usual, we have compared the irradiance pattern obtained by raytracing with, in this case, the irradiance pattern obtained by using the Lorentz geometry technique explained in this section. Figure 4.33 shows the comparison between the irradiance patterns.

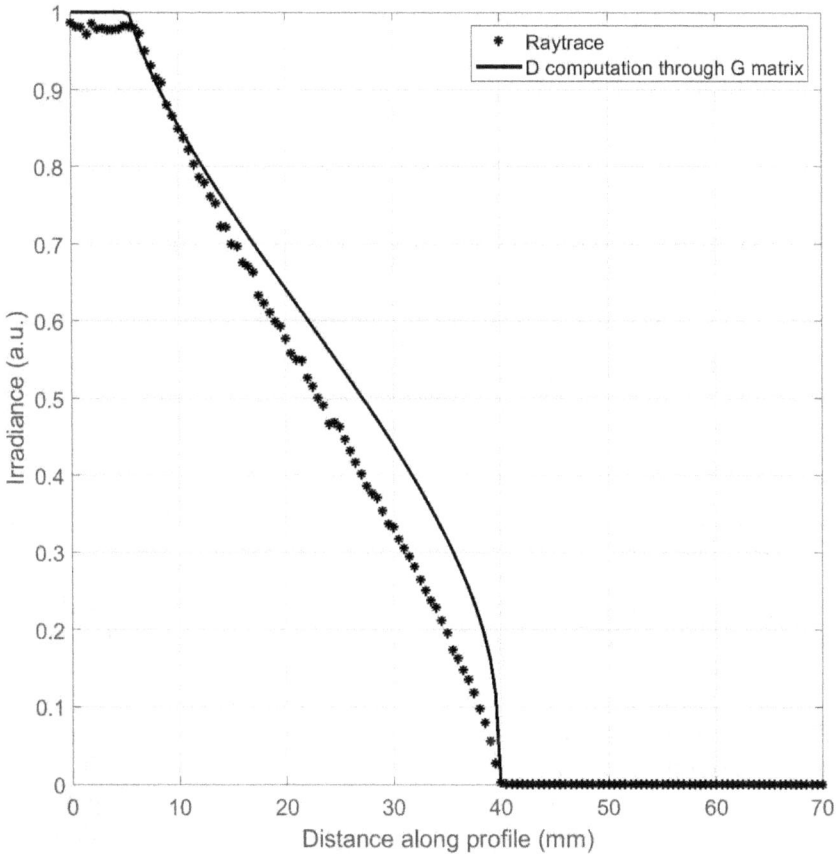

FIGURE 4.33: Comparison of irradiance profiles obtained by raytracing and by Lorentz geometry for the reflective system of fig. 4.32 with the detector at $z = 450\,mm$

4.8 $\vec{D}$ through Inhomogeneous Refractive Medium, Curved Cones of Edge Rays

In this section we are going to introduce techniques to obtain $\vec{D}$ inside inhomogeneous medium. As we have done in previous sections for homogeneous media, and in chapter 3 for free propagation, we are going to introduce two techniques to obtain $\vec{D}$: first, using the contour integral with special attention to eq. 4.11, and second, using Lorentz geometry, both of them based in the concept of the cone of edge rays. The main difference between homogeneous and inhomogeneous media is that the rays are straight lines in homogeneous

Inhomogeneous media

$n(x, y, z)$

Z

$\vec{D}$

Curved cone of edge rays

Y

X Lambertian source

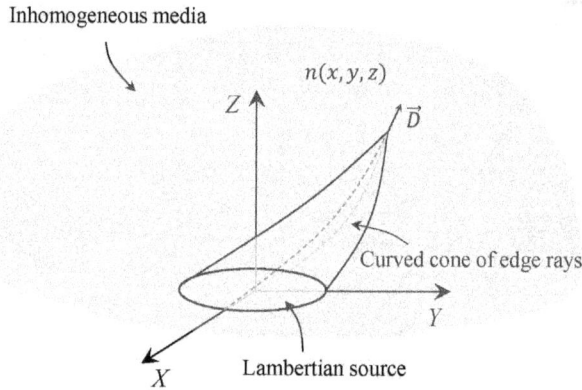

FIGURE 4.34: Curved cone of edge rays in inhomogenous refractive media

media and curved lines in inhomogeneous media, and the result is that the cone of edge rays in inhomogeneous media will be curved, fig. 4.34, and more difficult to analyze from a mathematical point of view.

As we have seen in the previous section and in chapter 3, eq. 4.11 provides a more simplified and accurate method of computing $\vec{D}$. For refractive/reflexive optical media we need the aid of Fermat's principle to obtain the refractive/reflexive cone of edge rays. The technique to compute $\vec{D}$ in inhomogenous media is based on the fact that we can always divide the inhomogeneous media in small volumes which can be considered homogeneous. For a spherical gradient index we can divide the space in small spherical layers with a constant refractive index; for an axial gradient index we can divide the space in small plane layers with a constant refractive index [55]. The smaller the layers, the higher the precision. Now using Fermat's principle it is possible to obtain the curved cone of edge rays in multilayer inhomogeneous media, fig. 4.35. To compute $\vec{D}$ at point P using eq. 4.11 we need the cone of edge rays produced at the layer where point P is located.

The application of Lorentz geometry to inhomogeneous media requires more sophisticated mathematical development. Gutierrez [44] studied the problem and states that every lightlike geodesic of the Lorentz metric must be a geodesic of the metric $n^2\delta$, where n is the refractive index function. The basic set of equations that must be fulfilled in inhomogeneous media is:

$$\left. \begin{array}{l} \ddot{\gamma}^k(t) + [\dot{\gamma}(t)]^k \Gamma^k[\gamma(t)]\dot{\gamma}(t) = 0 \\ \dot{\gamma}^k(t)G[\gamma(t)]\dot{\gamma}(t) = 0 \end{array} \right\} \Rightarrow \ddot{\gamma}^k(t) + \dot{\gamma}(t)\bar{\Gamma}^k[\gamma(t)]\dot{\gamma}(t) = 0, \quad (4.19)$$

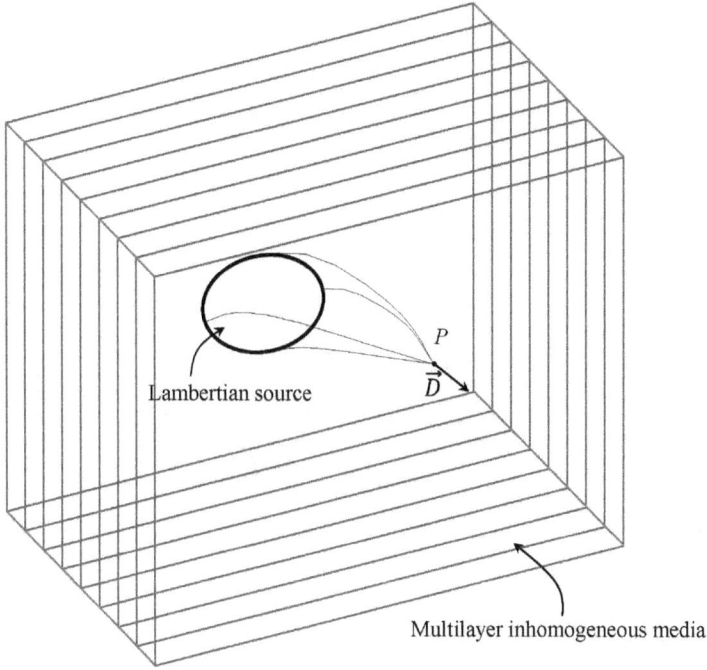

FIGURE 4.35: Sketch to compute $\vec{D}$ in inhomogeneous refractive media using eq. 4.11

where $\bar{\Gamma}_{ij}^{k}$ is the Christofel symbol of the metric $n^2\delta$ and Γ_{ij}^{k} is the Christofel symbol of the Lorentz metric. Note the difference between eq. 4.19 and eq. 3.90. As in section 3.9, eq. 4.19 provides a system of 18 first order partial differential equations; the solutions of this system of PDE provide the coefficients g_{ij} of the matrix G, from which it is possible to compute $\vec{D}$ at any point in space.

5

Thermodynamic Basis of the Irradiance Vector

5.1 Introduction

In section 1.2, we analyzed the ellipse-sphere paradox, which states a thermodynamic limit to the existence of point sources; the laws of thermodynamics place a theoretical limit also on the concentration ratio of nonimaging devices [56]. In this chapter we are interested in applying the laws of thermodynamics to determine performance limits for nonimaging optical systems. We consider the geometrical optics point of view through vector $\vec{D}$ and show that the first and second laws of thermodynamics, which are independent postulates, can be manifested through vector $\vec{D}$. In chapter 1 we studied the relation of $\vec{D}$ with the conservation of energy, and in the next section we will relate $\vec{D}$ to entropy. We will attempt to link the key concept of nonimaging optics, étendue, with the radiative heat transfer concept of view factor, which may be more familiar to some readers.

5.2 Relations between $\vec{D}$ and Thermodynamic Variables

In this section we are going to present and analyze some basic relations between $\vec{D}$ and thermodynamics variables like temperature, entropy and pressure. First we start with Stefan-Boltzmann law in Chapter 1 we stated that the emittance from a blackbody surface and irradiance upon a surface are connected by $\vec{D}$. Let us consider a blackbody source, with the total flux emitted as $\sigma a T^4$, where a is the surface of the blackbody and σ is the Stefan-Boltzmann constant. We can define a closed surface A, fig. 5.1, for which

$$\oint \vec{D} \cdot d\vec{A} = \sigma a T^4. \tag{5.1}$$

DOI: 10.1201/9780367551605-5

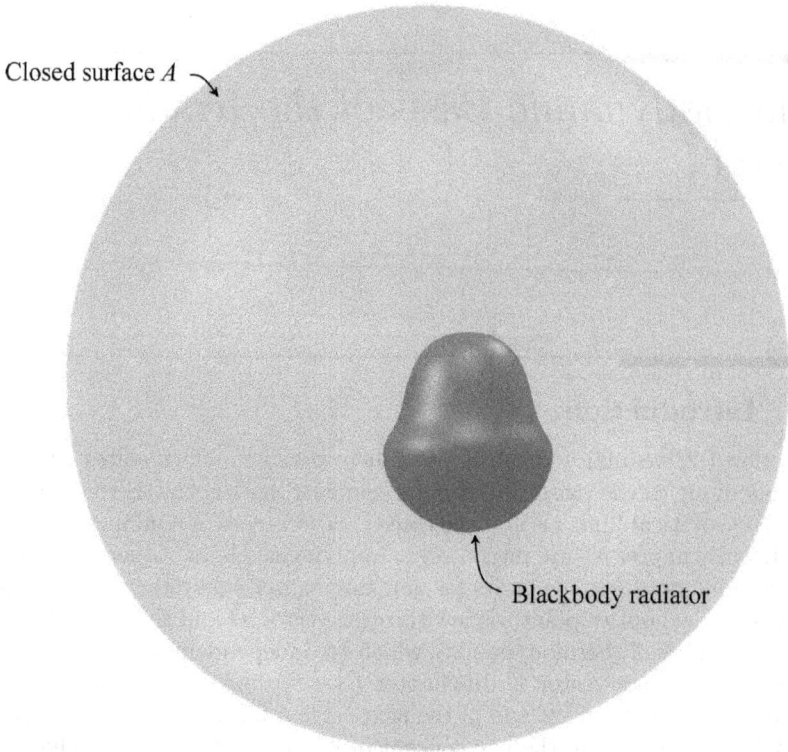

FIGURE 5.1: Closed surface integral around a blackbody source

We can make a step forward using Gauss's theorem in eq. 5.1,

$$\int \nabla \cdot \vec{D} dv = \sigma a T^4, \tag{5.2}$$

where v is the volume enclosed by the surface S. Equation 5.2 states that in the presence of a source the $div\vec{D}$ is not zero as expected, but also relates $\vec{D}$ to temperature . We will return to this concept in the next section.

Now we are going to present a second thermodynamic relation, in this case between $\vec{D}$ and entropy S. From statistical physics it is well known that the entropy is the logarithm of the phase space volume $\Delta p \Delta q$ [57]; in optics the relation between étendue U and phase space [51] is well established. So we can state that

$$S = k \log U, \tag{5.3}$$

where S is the entropy and k is Boltzmann's constant, considering the definitions of $\vec{J}$ and eq. 1.13, and its relation to étendue defined in eq. 1.8 and with $\vec{D}$ defined in eq. 1.16, it is possible to write

$$S = k \log \left(\frac{1}{B} \int \vec{D} \cdot d\vec{A} \right),$$ (5.4)

which relates the entropy S to the irradiance vector $\vec{D}$. This expression has been used in the design of fluorescent planar concentrators [58][59].

Another interesting thermodynamics relation comes from the Stefan-Boltzmann law. Let us take a volumetric blackbody radiator which emits uniformly in all directions. The brightness of the blackbody is

$$B = \frac{\sigma T^4}{4\pi}.$$ (5.5)

This relation states that for a Lambertian radiator we can replace the brightness with temperature; this can be done in the previous chapters of this book.

5.3 Thermodynamic Origin of Nonimaging Optics

Now that we have briefly introduced basic relations between $\vec{D}$ and thermodynamic variables we are going to study the connection between nonimaging optics and thermodynamics [60]. To connect thermodynamics with real world applications, we will demonstrate the underlying thought process of nonimaging optics with a realistic problem: concentrator designs. In order to simplify the model, we will assume that the designing process is purely geometric. In other words, effects such as wavelength shifts and polarization, which impact the thermodynamics of the system, are not our concern. For the readers who are interested in a more general analysis of such effects, which have to be considered in the design of luminescent concentrators, for example, please refer to [61]. Compared to imaging optics, a nonimaging concentrator is concerned with transferring energy from the source to the absorber (sink) within the theoretical limit of thermodynamics. Instead of designing an optical system based on fine tuning and engineering light rays, we first need to look at the maximum concentration ratio allowed by the fundamental laws of physics. To do so, we must enter the regime of geometrical optics and probability.

The general setup of a concentrating system requires three components, as shown in fig. 5.2: the source of radiation, the aperture of the concentrator, and the absorber (sink) of the radiation. We define the geometric concentration ratio as

$$C = \frac{Area\ of\ the\ aperture}{Area\ of\ the\ absorber}.$$ (5.6)

Radiation source

Aperture

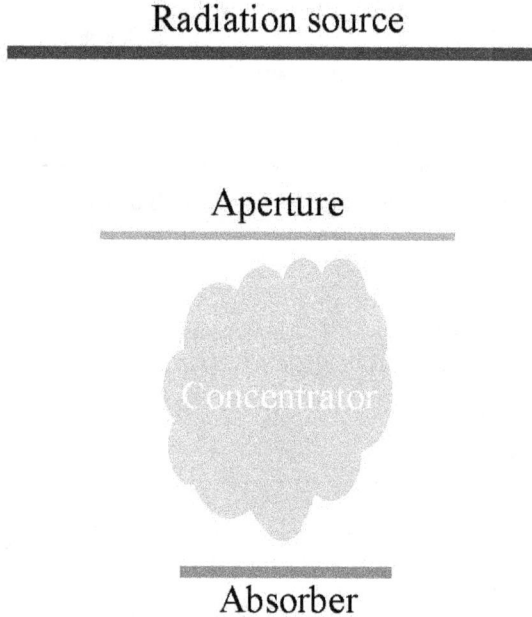

Absorber

FIGURE 5.2: Typical setup of a concentrator

To explore the thermodynamic limitations of a concentrator design using geometrical optics, we assume that the source and absorber are perfect Lambertian surfaces (blackbody). The probability of radiation leaving one surface and arriving at another surface is defined as

$$P_{ij} = \frac{radiation\ from\ i\ to\ j}{Total\ radiation\ emitted\ from\ i}. \qquad (5.7)$$

Now we shall discuss the first thermodynamic argument, which can be used independently of the second argument. The probability defined in eq. 5.7 can be readily utilized within the context of thermodynamics to provide the maximum concentration ratio of an optical device. For a configuration described in fig. 5.2, according to the Stefan–Boltzmann law, the radiation from a blackbody source to a blackbody absorber is

$$Q_{13} = A_1 \sigma T_1^4 P_{13}, \qquad (5.8)$$

where A_1 is the area of the source, σ is the Stefan–Boltzmann constant, and T is the temperature of the source. With the principle of reciprocity, which is a direct consequence of the second law of thermodynamics, we have

$$A_1 \sigma T^4 P_{13} = A_3 \sigma T^4 P_{31}, \quad A_1 P_{13} = A_3 P_{31}. \tag{5.9}$$

Equation 5.9 represents a fundamental concept: given two blackbodies at the same temperature, the radiative power from one to the other is equivalent. From here, we derive the radiative heat flux at the surface of the absorber

$$q_{13} = \frac{Q_{13}}{A_3} = \frac{A_1 P_{13}}{A_3} \sigma T_1^4 = P_{31}, \sigma T_1^4. \tag{5.10}$$

Because $P_{31} \leq 1$, q_{13} reaches a maximum radiative flux equal to that of the source, which is consistent with the second law of thermodynamics

$$q_{13} = \sigma T_1^4 \text{ if and only if } P_{31} = 1. \tag{5.11}$$

Now we can discuss the second thermodynamic argument. Ideally, a concentrator design has a geometrical optical efficiency equal to one. In other words, for a configuration shown in fig. 5.2, the energy from the radiation source passing through the aperture will all arrive at the absorber (for nonideal systems, refer to [62])

$$Q_{12} = Q_{13}, \quad A_1 \sigma T^4 P_{12} = A_1 \sigma T^4 P_{31}, \quad A_1 P_{12} = A_1 P_{31}. \tag{5.12}$$

Equation 5.12 provides the limit of concentration given the additional requirement that the optical efficiency of the device is also ideal. Such a requirement is not always enforced in the real world application; theoretically, however, providing a device with the ideal design is at least an option to be considered. Using the principle of reciprocity once more, we have

$$A_1 P_{12} = A_1 P_{31}. \tag{5.13}$$

Combining Eqs. 5.9, 5.12, and 5.13, we conclude for concentration ratio

$$C = \frac{A_2}{A_3} = \frac{P_{13}}{P_{21}} \leq \frac{1}{P_{21}}. \tag{5.14}$$

Equation 5.14 demonstrates the design principle of an ideal nonimaging concentrator. Impose the following requirements:

1. Maximize the concentrated flux at the absorber.
2. Maximize the optical efficiency of the concentrator.

Then the concentration ratio of such a device is $\frac{1}{P_{21}}$. Equation 5.14 also shows that such a design is achieved when $P_{13} = 1$, according to eq. 5.11. The thermodynamic approach of maximizing the absorber temperature such that it is equivalent to the radiation source, or $P_{31} = 1$, serves as the underlying principle for most nonimaging designs.

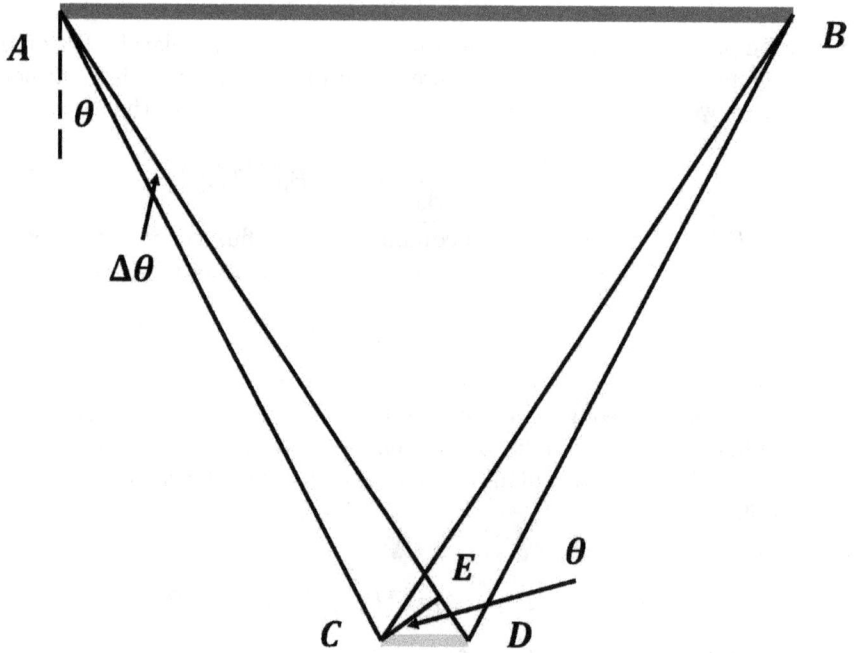

FIGURE 5.3: Hottel's strings can be used to solve for P_{21}

Hoyt Hottel, an MIT engineer working on the theory of furnaces [63], showed a convenient method for calculating radiation transfer between walls in a furnace using strings. In order to calculate the P_{21} from eq. 5.14, we use Hottel's strings on the radiation source 1 and aperture 2, as shown in fig. 5.3:

$$P_{21} = \frac{(\overline{AD} + \overline{BC}) - (\overline{AC} + \overline{BD})}{2\overline{CD}}. \tag{5.15}$$

As the source AB approaches infinity, $\Delta\theta$ approaches 0 and $\overline{AC} = \overline{AE}$. If the setup is kept symmetric, i.e., $\overline{AD} = \overline{BC}$ and $\overline{AC} = \overline{BD}$, then

$$P_{21} = \frac{2(\overline{AD} - \overline{AC})}{2\overline{CD}} = \frac{\overline{DE}}{\overline{CD}} = \sin\theta, \quad C_{max} = \frac{1}{P_{12}} = \frac{1}{\sin\theta}. \tag{5.16}$$

Equation 5.16 gives the maximum ratio of concentration within the framework of thermodynamics for an infinitely far away radiation source in 2D. One can easily generalize this limit to 3D:

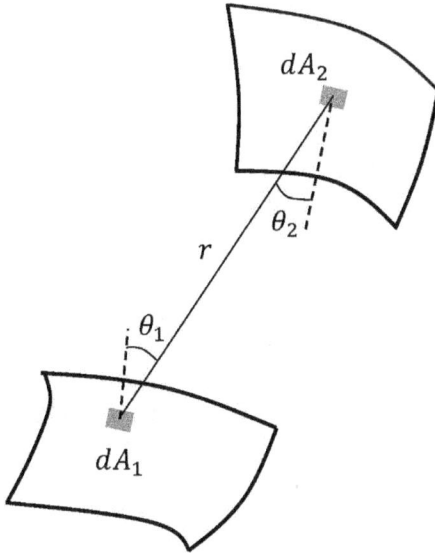

FIGURE 5.4: Radiative heat transfer between areas 1 and 2

$$C_{max} = \frac{1}{\sin^2 \theta}. \tag{5.17}$$

Now we are going to explain the concept of étendue based on conventional radiative heat transfer. This might be particularly helpful for readers who are familiar with such a background, and it can provide the reader with a clearer view of the connections between heat transfer and nonimaging optics. In the context of radiative heat transfer, the geometric setup is described using view factors [54]. Similar to [39], a probability concept can be established to describe the geometric capability for a radiative heat transfer device, even though the source and absorber may not have a direct view of each other:

$$P_{12} = \frac{1}{\pi A_1} \int \int_{A_1 A_2} \frac{\cos \beta_1 \cos \beta_2}{r^2} dA_1 dA_2. \tag{5.18}$$

Here dA_1, dA_2 are two infinitely small areas on the surfaces of two objects undergoing radiative heat transfer (fig. 5.4). Variables β_1 and β_2 are the angles formed between the direction connecting them and the norms of the specific surface, and r is the distance between them. Because we are in a geometric optical setup, some of the rays from the source of radiation to the sink can be reflected or refracted. Probability P instead of view factor F is used to include these rays. In other words, compared to the definition of view factor F, which only describes geometric configurations without considering optical devices, probability P is a more general description of radiative heat transfer which includes ray paths that connect the source and the absorber through optical devices such as mirrors and lenses. From eq. 5.18, the étendue ([24],[51]) is

$$E_{12} = \int\int_{A_1 A_2} \frac{\cos\beta_1 \cos\beta_2}{r^2} dA_1 dA_2. \tag{5.19}$$

Comparing eqs. 5.18 and 5.19, we conclude

$$E_{12} = \pi A_1 P_{12}. \tag{5.20}$$

This result connects the étendue with view factor analysis , which is an important concept in radiative heat transfer studies, and its tabulated analytical results are widely available in the literature [54]. Notice that for nonimaging optical systems designed in 2D, $E_{12} = 2A_1 P_{12}$ due to the integral in eqs. 5.18 and 5.19 being limited in 2D. The second law of thermodynamics forbids a colder body to transfer heat to a hotter body. Therefore (also from eq. 5.18), this can be described from the étendue perspective as

$$E_{12} = \pi A_1 P_{12} = \pi A_2 P_{21} = E_{21}. \tag{5.21}$$

Or, for any two surfaces we can conclude that

$$E_{ij} = \pi A_i P_{ij} = \pi A_j P_{ji} = E_{ji}. \tag{5.22}$$

We can implement eq. 5.19 with eqs. 5.12 and 5.13 and conclude that the shorthand expression of the two important thermodynamic arguments for nonimaging optics is [35]

$$E_{12} = E_{13} = \pi A_3. \tag{5.23}$$

Equation 5.23 shows that the ideal nonimaging optical designs are geometrically étendue matching devices. If we treat étendue as the ensemble of all the rays connecting two objects (such as radiation source, aperture, or absorber), then we can observe that for the nonimaging geometric optical design that achieved maximum concentration ratio with theoretical 100% optical efficiency, eq. 5.23 is required from a thermodynamic point of view.

5.4   Blackbody Radiation in a Cavity and the Radiation Pressure

In this section we are going to study the blackbody radiation in a enclosed cavity in equilibrium, fig. 5.5. This question, sometimes called the photon gas problem, has been well studied from the point of view of statistical physics

Blackbody cavity in equilibrium at T

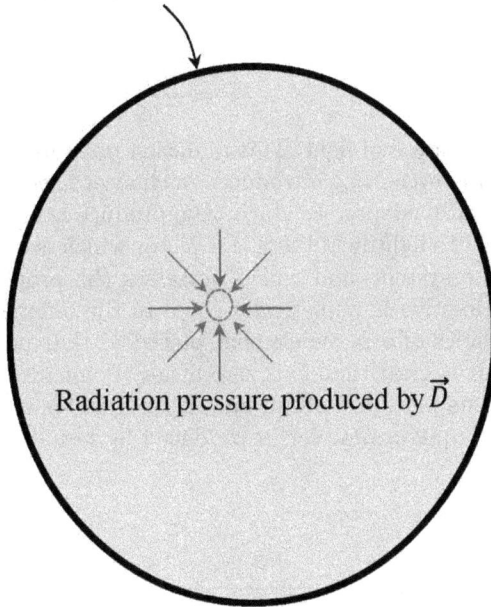

Radiation pressure produced by $\vec{D}$

FIGURE 5.5: Blackbody cavity in thermodynamic equilibrium at temperature T

[57][64]. We are going to use well-known results of statistical physics, and we are going to connect them with $\vec{D}$. The equilibrium mechanism in 98 gas in the cavity consists of the absorption and emission of photons by the matter of the cavity. This produces an important result: the number of photons in the cavity N is variable. Another important property of photon gas in equilibrium is that the interaction between the photons themselves may be regarded as completely absent. The thermodynamic properties of blackbody radiation in equilibrium in a cavity can be obtained from Planck's radiation law

$$dE_\omega = \frac{V\hbar}{\pi^2 c^3} \frac{\omega^3 d\omega}{e^{\frac{\hbar\omega}{kT}} - 1}, \tag{5.24}$$

where the energy of the photon is $\epsilon = \hbar\omega$, V is the volume of the cavity and k is the Boltzmann constant. By integration of eq. 5.24 over ω we can obtain the Stefan-Boltzmann law, eq. 1.5. It is possible to define the pressure of the photon gas in the blackbody cavity, from a statistical physics point of view it can be given as [57]

$$PV = \frac{1}{3}E \tag{5.25}$$

and

$$P = \frac{4\sigma T^4}{3c}, \tag{5.26}$$

where c is the speed of light. This radiation pressure can be explained as radiation pressure force $\vec{F}_{rad}$, introduced in chapter 1, eq. 1.4. Using the definition of $\vec{D}$ provided in chapter 1, which we reproduce here for more comfort, "... at any point of the light field there is a vector which is independent of the choice of coordinate system, and which possesses the property that its projection upon any direction is numerically equal to the difference of scalar irradiance of the two sides of a plane element placed at that point and normal to that direction." It is clear that $\vec{D}$ is zero at any point inside the blackbody cavity in thermodynamic equilibrium, but the pressure is not zero, in a similar way to a fluid in equilibrium, and is produced by radiation force related to $\vec{D}$ by eq. 1.4.

5.5   Kirchhoff Radiation Law

In previous sections we have studied the properties of blackbody radiation sources and their relation to $\vec{D}$; now we are going to analyze a more fundamental property in radiation transfer, Kirchhoff's radiation law, and its relation to $\vec{D}$. It is well known that a blackbody radiation source absorbs all radiation incident on it. The monochromatic emissive power of a surface can be expressed by

$$e_\lambda = \epsilon_\lambda e_{b\lambda}, \tag{5.27}$$

where $e_{b\lambda}$ is the monochromatic emissive power of a blackbody surface, and we can define the monochromatic hemispherical emittance of the surface ϵ_λ; in a similar way we can define the monochromatic hemispherical absorptance of a surface α_λ [32]. An important relation between these two parameters can be obtained from Kirchhoff's law. The complete derivation of this law is in Planck's book [65]. Kirchhoff's law states that for a system in thermodynamic equilibrium, the monochromatic hemispherical emittance is equal to monochromatic hemispherical absorptance.

$$\epsilon_\lambda = \alpha_\lambda. \tag{5.28}$$

Equation 5.28 describes a fundamental law of nature; in fact, this relation also applies for nonequilibrium conditions. Kirchhoff's law states that ϵ_λ and α_λ

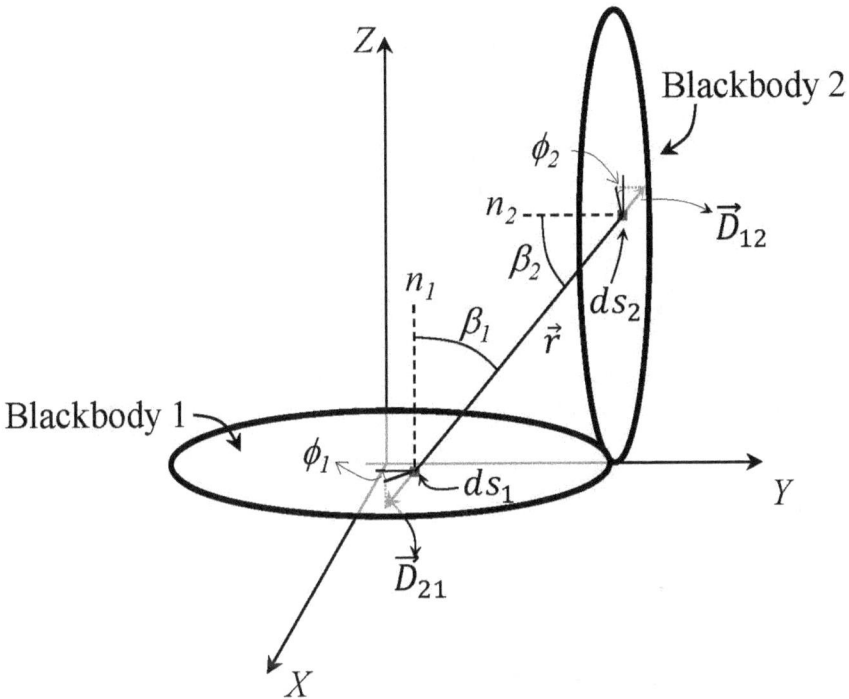

FIGURE 5.6: Interaction between two blackbodies in thermodynamic equilibrium

are the same. This is a fundamental property of the radiation source. For a blackbody, $\epsilon_\lambda = \alpha_\lambda = 1$. This fundamental law has relevant consequences in the analysis of radiation transfer between bodies. Remember that for a blackbody the emittance depends only on its temperature, which is a fundamental physical magnitude which characterizes the body in a similar way to mass or charge. To illustrate this idea we are going to study the interaction between two blackbodies in thermodynamic equilibrium. For simplicity we will analyze the configuration of fig. 5.6 with two circular blackbodies of radius R: blackbody one (BB_1) with its center at the origin of the reference system and blackbody two (BB_2) orthogonal to BB_1 and with its center in the ZY plane. We are going to evaluate the resultant $\vec{D}$ produced by BB_1 in BB_2, $\vec{D}_{12}$, and using eq. 5.28 connect it with the resultant vector produced by BB_2 in BB_1, $\vec{D}_{21}$. Starting with $\vec{D}_{12}$ and integrating eq. 1.3, we have

$$\vec{D}_{12} = \int_{s2} \int_{s1} \frac{B_1 \cos(\beta_1) \vec{u}_r}{r^2} ds_1 ds_2, \tag{5.29}$$

with ds_1 being the surface element of BB_1 and ds_2 the surface element of BB_2. Due to the symmetry of the fig. 5.6, $\vec{D}_{12}$ has only two components, D_{12y} and D_{12z}. D_{12y} can be easily computed:

$$D_{12y} = \int_{s2} \int_{s1} \frac{B_1 \cos \beta_1 \cos \beta_2}{r^2} ds_1 ds_2, \tag{5.30}$$

which is analogous to eq. 5.19. The D_{12z} component can also be obtained by

$$D_{12z} = \int_{s2} \int_{s1} \frac{B_1 \cos \beta_1 \sin \beta_2 \cos \phi_2}{r^2} ds_1 ds_2, \tag{5.31}$$

where ϕ_2 is the angle between the projection of $\vec{D}_{12}$ to XZ plane and Z axis. $\vec{D}_{21}$ has also D_{21y} and D_{21z}, but in this case

$$D_{21z} = \int_{s2} \int_{s1} \frac{B_2 \cos \beta_1 \cos \beta_2}{r^2} ds_1 ds_2, \tag{5.32}$$

and for D_{21y} component

$$D_{21y} = \int_{s2} \int_{s1} \frac{B_2 \cos \beta_2 \sin \beta_1 \cos \phi_1}{r^2} ds_1 ds_2. \tag{5.33}$$

Analogously ϕ_1 is the angle between the projection of $\vec{D}_{12}$ to XY plane and Y axis. From the geometry of fig. 5.6 we have

$$\cos \beta_1 = \sin \beta_2 \cos \phi_2 \tag{5.34}$$

and

$$\cos \beta_2 = \sin \beta_1 \cos \phi_1, \tag{5.35}$$

and eq. 5.31 and eq. 5.33 can be rewritten as

$$D_{12z} = \int_{s2} \int_{s1} \frac{B_1 \cos^2 \beta_1}{r^2} ds_1 ds_2 \tag{5.36}$$

and

$$D_{21y} = \int_{s2} \int_{s1} \frac{B_2 \cos^2 \beta_2}{r^2} ds_1 ds_2. \tag{5.37}$$

For the two blackbodies in thermodynamic equilibrium the brightness is the same:

$$B_{\lambda 1} = B_{\lambda 1} = 2\pi e_{b\lambda}. \tag{5.38}$$

Then by the geometry of fig. 5.6, $D_{12z} = D_{21y}$, and we can write the relation

$$\vec{D}_{12} = -\vec{D}_{21}. \tag{5.39}$$

Considering that no reflection is produced between them, only absorption such that the coefficient κ of eq. 1.4 only depends on α_λ, then

$$\alpha_{\lambda 2}\vec{D}_{12} = -\alpha_{\lambda 1}\vec{D}_{21}, \tag{5.40}$$

and it is possible to conclude that the radiation force from BB_1 to BB_2 is the same as the radiation force from BB_2 to BB_1, fig. 5.6:

$$\vec{F}_{rad\ 12} = -\vec{F}_{rad\ 21}. \tag{5.41}$$

Therefore it is possible to consider Kirchhoff's law, eq. 5.28, as an expression of the action-reaction principle .

6

Phase Space in Nonimaging Optics

6.1 Introduction to Phase Space in Nonimaging Optics

Phase space is the space in which the laws of statistical physics are developed. Statistical physics studies the properties of macroscopic systems, which means systems composed of a great number of particles or atoms; typically the number of particles is about the order of magnitude of Avogadro's number . The thermodynamic properties of macroscopic systems can be obtained easily using the laws of statistical physics [57] [64]. In the previous chapter we studied the relation between $\vec{D}$ and basic thermodynamic variables like temperature, pressure or entropy. Then a natural step forward appears to be the study of phase space properties of the nonimaging optics field, in which our particle will be a ray. In section 1.4 we introduced the concept of the optical momentum of a ray p_x and p_y. The space composed of the position of a ray q_i and the momentum of a ray (q_i, p_i) allows us a representation of the system in phase space. To illustrate this concept we are going to start our analysis with a simple geometrical optics example.

To understand the phase space of geometric optics, we will use the simple example shown in fig. 6.1. A source at $z = 0$ emits light rays which subtend a half angle of 30 deg. Here we define the optical axis as the z axis. Figure 6.1a shows how we can discretize both the position x and the angle of such rays [19]. We pick four equally spaced x positions and use seven equally spaced angles to represent all the rays from the screen. In fig. 6.1b we plot these rays as points according to their x positions and their directions as they intercept a screen at various z positions, which are represented as the vertical lines in fig. 6.1. However, instead of plotting the ray's direction as angles, we plot the direction according to the sine of the incident angle, or $p_x = n \sin \theta$; p_x is also called optical momentum under Hamiltonian optics [66], which is equal to directional cosine when the refractive index is 1. Because such angles are between 30 and –30 deg, the representations of the rays are also between 0.5 and –0.5 on the p_x axis. The screen starts at $z = 0$, or when the screen is right against the source, the "x" markers occupy a rectangular shape. The four corners of this rectangle represent the four extreme rays $(-5, -sin(30))$, $(-5, sin(30))$, $(5, sin(30))$ and $(5, -sin(30))$. Next, we move the screen to

DOI: 10.1201/9780367551605-6

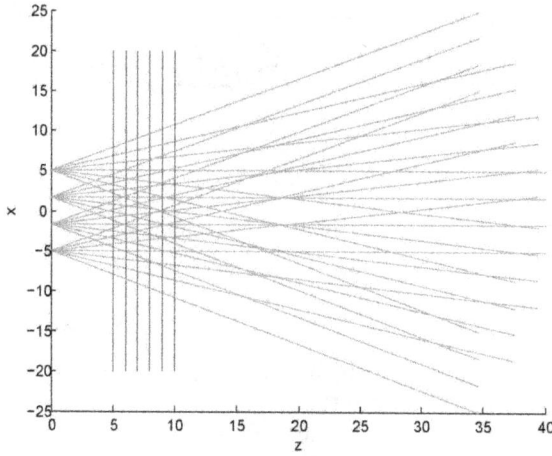

a)

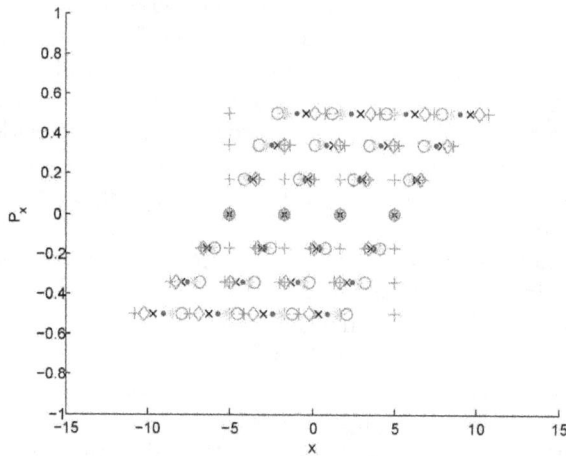

b)

FIGURE 6.1: Discrete representation of phase space

position $z = 5$ to intercept these rays, and we plot the rays again in phase space with the "o" markers.

Now the extreme rays with positive directions drift toward the positive x direction, and the extreme rays with negative directions drift toward the negative x direction. The rays incident on the screen occupy the shape of a parallelogram. As the screen takes on further positions such as $z = 6, 7, 8, 9, 10$

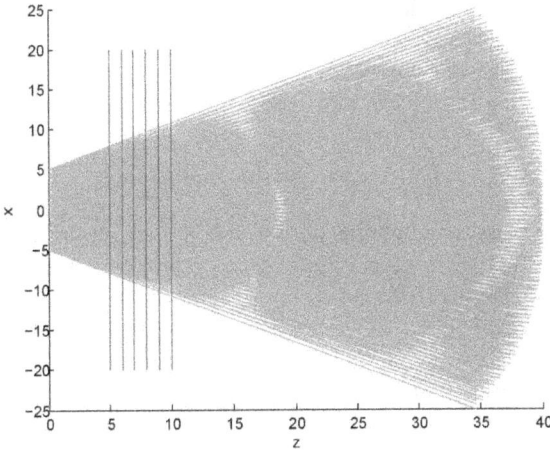

a)

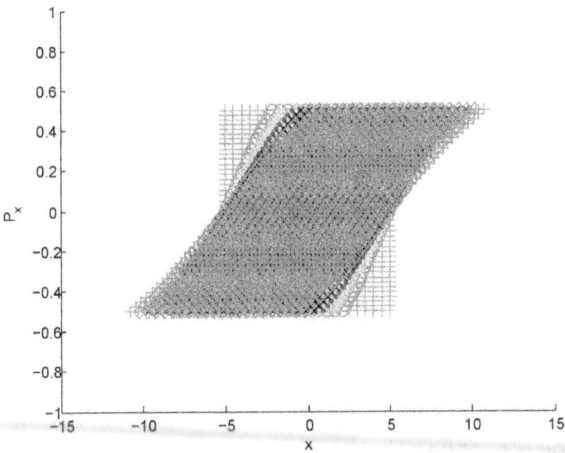

b)

FIGURE 6.2: Dense representation of phase space

we see that the parallelogram becomes more skewed. This can be seen more clearly as we increase the density of the rays (fig. 6.2). In fig. 6.2 it is shown that such parallelograms occupy the same area. Without giving further detailed proof [54], we state that without any loss or generation of rays, the phase space volume is always conserved.

This conservation rule can be analyzed using the concept of étendue in the phase space framework. We observe in fig. 6.2b that the area in the phase space occupied by the rays remains the same. This area corresponds to the value of étendue, or the spread of the rays in both positions and directions. We defined étendue in eq. 1.8, and we can express this as

$$\int dp_x dx = \text{étendue} = \text{constant}. \qquad (6.1)$$

The conservation of étendue can be derived from Hamiltonian optics ; other forms of proofs have also been offered ([24], Appendix A.2, A.3). In the next section we also offer a proof due to Fermi [15]. For a system described in fig. 6.2, it can be visualized as follows. First, we choose an arbitrary position on the z axis. The rays of an optical system that intersect a screen at this position (fig. 6.2a) can be represented, one to one, as points in a phase space with x, p_x coordinates (fig. 6.2b). As we move the screen along the z axis, the boundary of such a phase space representation of all the rays in the optical system will also shift its shape. However, the area, or étendue, within this boundary will remain constant. In an analogy, if we treat z, or the position of the screen, as time, then as we move away from the screen (changing the time z), the étendue of the system is like the volume of a noncompressible fluid. As the fluid takes on different shapes, its volume will remain the same.

Now consider the concentrator design in air ($n = 1$), with a specified profile of aperture phase space as shown in fig. 6.3a. All the rays coming into the aperture ($-a, a$) subtend the half acceptance angle (notice fig. 6.1 simply consists of the reversed rays, with $\theta = 30$, $a = 5$), which is typical for a far-away radiation source; then we can draw its phase space representation in fig. 6.3b. To fit the phase space volume (area in 2D), or étendue, into the smallest absorber area, the following limitation exists: the phase space volume cannot extend above $L = 1$ or below $L = -1$. This limitation exists because directional cosine exceeding 1 is nonphysical. Thus we arrive at a rectangular phase space volume, as shown in fig. 6.4a.

Consider the following proof. If the phase space volume of the absorber takes on a shape other than the rectangular area with vertical boundaries, as shown in fig. 6.4b, the absorber phase space area is the parallelogram outlined by the black dotted–dashed lines. Then we can always find a new boundary with vertical line $x = x_0$ that allows part of the phase space area outside the new boundary to be folded inside, resulting in a smaller absorber area, where the new physical size of the absorber starts with $x = x_0$. In other words, having a nonvertical boundary for phase space volume means that certain small areas on the absorber are under-utilized. Or, some of the angles for such small areas are not being used to receive incoming radiation. In such a situation, one ends up wasting the phase space volume that the absorber has

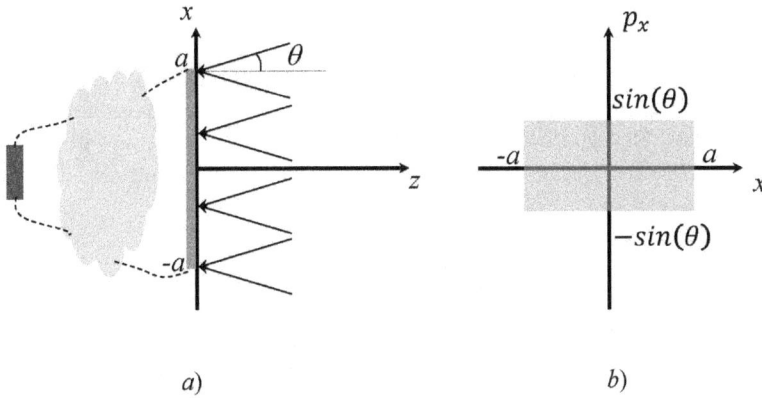

FIGURE 6.3: Discrete representation of phase space

for accepting incoming light. A quick calculation shows that the size of the new absorber is $2a \sin \theta$; therefore, the ideal concentration ratio is

$$C_{max} = \frac{1}{\sin \theta}. \tag{6.2}$$

It is easy to pictorially describe the phase space concept of a 2D system; however, this is not the case for a 3D system. In a 2D system we can use x and directional p_x to uniquely identify a ray, as we have shown in fig. 6.3. In order to describe the 3D rays in the phase space geometrically, we need four unique parameters instead of two: x, y, directional p_x , and directional p_y. This

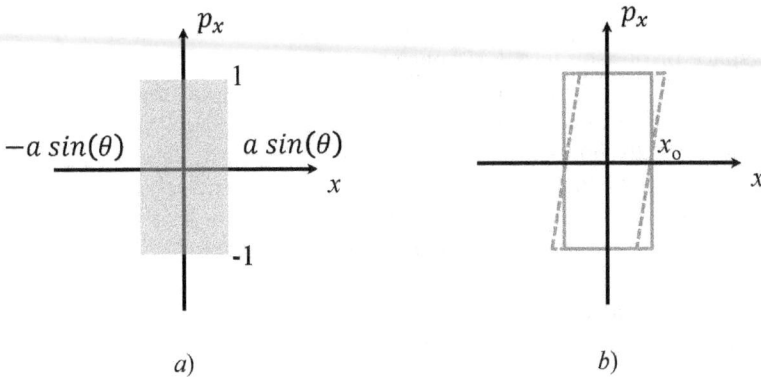

FIGURE 6.4: Dense representation of phase space

four-dimensional space is no longer readily intuitive for us, but we can still observe its properties in its projections. When we project it into the x, y space it takes on the unit of $\frac{watts}{m^2}$, which is represented as the typical irradiance for a screen, but if we instead project into the directional cosine spaces p_x, p_y it becomes $\frac{watts}{sr}$. Later on we will discuss how an ideal nonimaging system behaves under such a projection. First, we can expand the same phase space representation of étendue into 3D configurations. The étendue is defined as eq. 1.8:

$$\int dp_x dp_y dx dy = \text{étendue} = U. \tag{6.3}$$

Here the phase space volume is calculated based on the four-dimensional coordinates x, y, p_x, p_y. For any point x, y, the set of rays coming from the source looks the same:

$$dU = dp_x dp_y dx dy. \tag{6.4}$$

When the aperture is illuminated with identical pencil rays—in other words, p_x, p_y are independent according to x, y, fig. 6.5a—

$$U = \int dp_x dp_y dx dy = \int dp_x dp_y \int dx dy = A \int dp_x dp_y = n^2 A \int dL dM, \tag{6.5}$$

where A is the area of the aperture. L and M are directional cosine x and directional cosine y, respectively. As we mentioned before, although it is hard to visualize such a four-dimensional volume, for certain configurations such a calculation can be greatly simplified. For example, given an aperture receiving light from an infinitely far away source, we can pick any point (x, y) on the receiver, and the set of rays going through a small area $dx dy$ will not be dependent on the (x, y) position, fig. 6.5a. We can visualize the 4D phase space of such an aperture by projecting it on the 2D LM or $(p_x p_y)$ space. The phase space projection into LM space in such an example will be of a circular area of radius $\sin \theta$, as shown in fig. 6.5b. The phase space volume is therefore $nA \sin^2 \theta$. We can also calculate the $\int dL dM$ using the unit sphere method [54]. It is crucially important to understand that the directional cosine space is different from the 2D phase space. Although they are both two-dimensional, one represents the directional x and directional y (L and M, or $p_x p_y$), which is a projection of the full four-dimensional phase space. The other is the full representation of x and p_x under a 2D configuration.

We have provided two ways of arriving at the same limit of concentration. With the thermodynamic argument eq. 5.16 and the phase space argument eq. 6.2, we reached the same conclusion. Equation eq. 5.16 provides the limit of concentration under simple but fundamental thermodynamic assumptions.

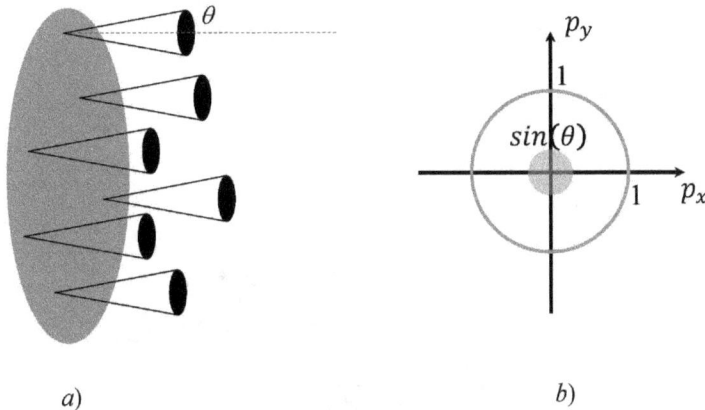

FIGURE 6.5: a) 3D configuration of light coming into aperture A subtending angle θ. b) The $\frac{L}{M}$ space representation of the incoming rays

With eq. 6.2 we see how étendue conservation in phase space will provide clues for the designs for edge rays; a more robust argument can be found in [61]. Although such thermodynamic understanding of nonimaging optics does not offer design methods to directly generate a nonimaging concentrator, it does provide a theoretical limit and some intuition for the designing process. It also serves as a guideline on how nonimaging optical systems can be designed, as shown below.

One of the first nonimaging optical designs to be invented was the compound parabolic concentrator (CPC) , which we look at now. Here the aperture is between two groups of edge rays that represent the two extreme angles that the source subtends. The most rudimentary way of designing a concentrator for such a setup is to use two straight line reflectors, as shown in fig. 6.6. This, indeed, was the first effort employed that led to the design of the CPC. However, it will result in a poor concentration ratio. The evolution of such a design into the CPC has been well documented [51]. Here instead, we use the insights provided by eq. 5.14 to review the result of this designing process to give another perspective of how nonimaging optics is related to thermodynamics. We require that:

1. $P_{31} = 1$, or all rays from the absorber back trace to the angle between the two incident directions

2. $P_{12} = P_{13}$, or all rays between the two extreme directions will arrive at the absorber.

Then the straightforward attempt is to map the group of edge rays from direction 1 to absorber edge $B\prime$ and the group of edge rays from direction 2 to

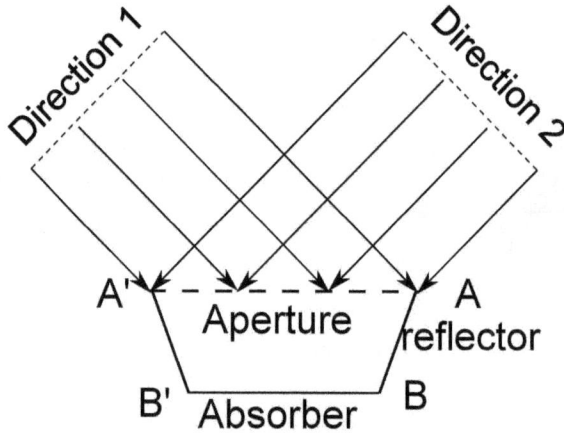

FIGURE 6.6: Using simple straight line reflectors to form the concentrator

absorber edge B. This can be shown as the following. To understand P_{31}, we look at the extreme rays, namely, rays from points B and B'. To prove that for $P_{31} = 1$ requires that all rays from B and B' should be mapped into extreme ray directions θ and $-\theta$, we use this following proof of contradiction. If some of the rays from B or B' are mapped into angles larger than θ, say, at θ_0, then we can find their neighboring point B_0, where the small ray cluster emitted by BB_0 at θ_0 will be outside of the source 1. Similarly, if some of the rays from B or B' are mapped into angles smaller than θ, then some rays coming into the aperture will not be mapped between B and B'. In such a case we would not be able to guarantee $P_{12} = P_{13}$. By limiting all the rays from points B and B' to be mapped to extreme angles, we ensure both conditions that are implied by eq. 5.14. Mapping these two groups of parallel rays into two points will result in two parabolic curves, fig. 6.7. We determine the aperture AA' when the tangent of parabola becomes vertical, because under this condition the aperture of the concentrator will be at a maximum. This, however, is only one way of designing optics that satisfy both thermodynamic requirements of eq. 5.13.

Now we prove that the concentrator constructed this way does achieve the ideal concentration ratio. Using the geometric property of the parabola in fig. 6.7 we have

$$CA + AB' = A'B + BB' \quad \text{and} \quad A'B = AB', \tag{6.6}$$

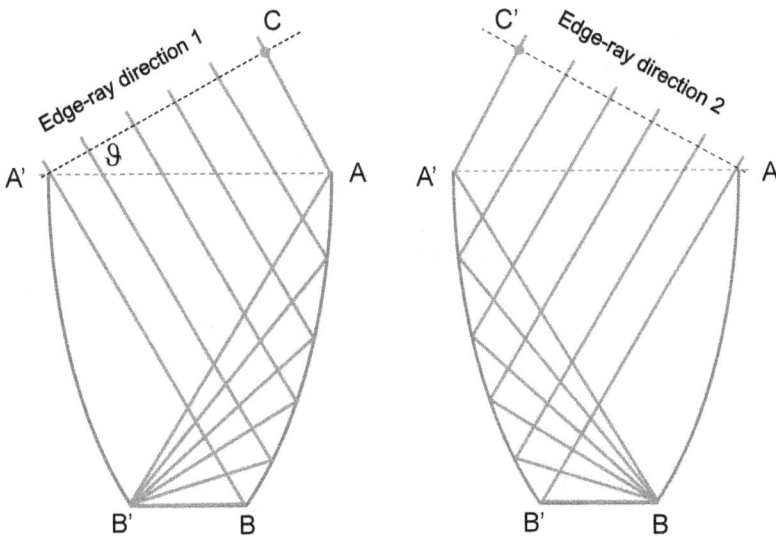

FIGURE 6.7: Mapping points to edge rays

therefore

$$CA = BB'. \tag{6.7}$$

Utilizing the property of right triangle $AA'C$

$$AC = A'A \sin \theta \tag{6.8}$$

gives us the concentration equation

$$\frac{A'A}{B'B} = \frac{1}{\sin \theta}. \tag{6.9}$$

Phase space of a plane in the optical device can be plotted based on the position and direction of a light ray. Here we use the CPC design from the previous section as an example. First, we examine the phase space representation of the aperture plane $A'A$. At point A', the two extreme rays both intersect with the aperture at the same x coordinate; therefore, their representation on the phase space coordinate has the same value on the x axis. In phase space, instead of plotting the ray's direction as the incident angle, we plot the direction according to the sine of the incident angle, or $p_x = n \sin \theta$, where n is the index of refraction. We also call p_x the optical momentum or, when the refractive index is 1, directional cosine. The incident angle of ray 1 is θ, while ray 2 is at $-\theta$. Therefore, they are plotted as points 1 and 2 in the phase space. The same can be done for rays 3 and 4, which are represented as points 3 and

4 in the phase space plot. All the other rays incident at point A' have the same x coordinates, with different values for p_x between $\sin\theta$ and $-\sin\theta$. As a result, they will form a line between points 1 and 2. As we plot these phase space points one by one based on their position of x and directional cosine p_x, we find that this cloud of points occupies an area shaped as a rectangle. The same plot can be created for the absorber. In this case the extreme rays will intersect with the absorber at angles of 90 deg; therefore, their p_x coordinates correspond to 1 and -1, corresponding to the limits we saw for L in fig. 6.3b. For an optical device without light generation or loss, its phase space volume is conserved [23]. In a 2D configuration, as shown in fig. 6.8, the phase space volume, represented by the area of the rectangles, is conserved between the aperture and the absorber plane,

$$A'A\sin\theta = B'B\cdot 1, \;\; C = \frac{A'A}{B'B} = \frac{1}{\sin\theta}, \tag{6.10}$$

which shows that, from the phase space conservation perspective, the CPC is achieving the ideal concentration limit.

It is well known that nonimaging optics only needs to map rays from the boundary to the source to the boundary to the receiver. The edge ray principle requires that the rays traveling at extreme directions and or emanating from extreme positions at the aperture must be directed to the extreme

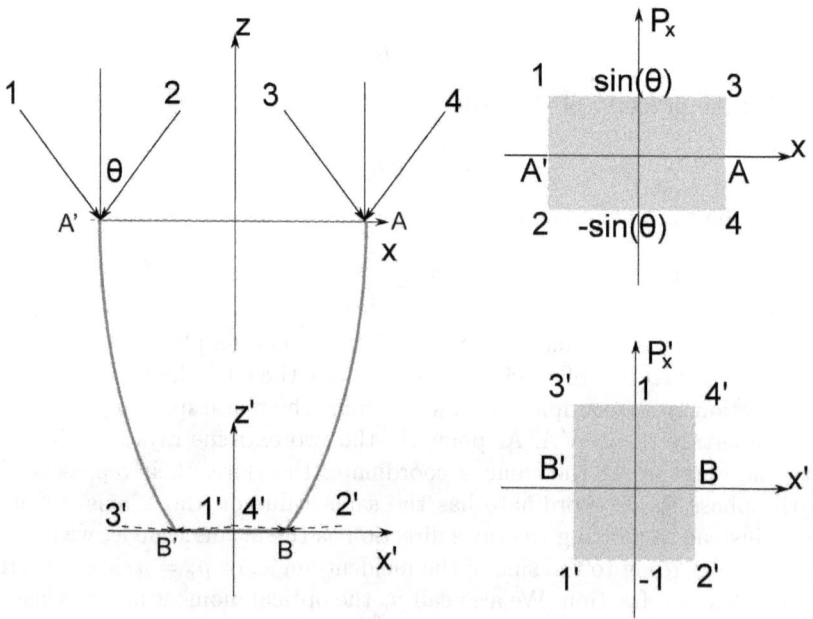

FIGURE 6.8: Phase space of a CPC

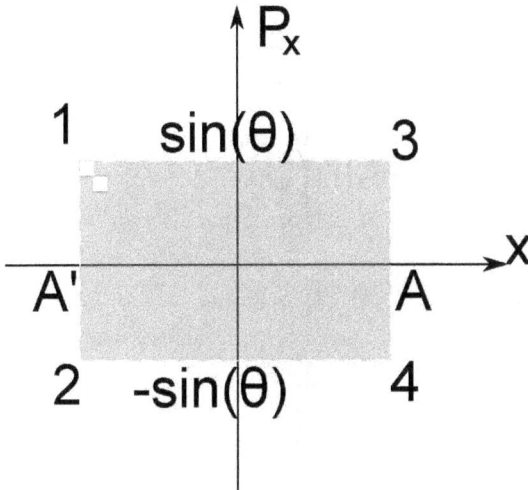

FIGURE 6.9: Ray 1 cannot interchange its position with a ray that is not originally inside the boundary

directions and or extreme positions at the absorber. Extreme is meant to refer to the largest angles accepted by a system and or the edge-most positions located on an aperture or absorber. We call such rays the edge rays. When plotted in phase space, they are represented by the boundaries of the phase space volume (area for 2D). We use the phase space representations of an optical device to prove this concept, which is similar to the Liouville theorem in classical mechanics [67]. As shown in fig. 6.9, it is impossible for a ray to retire from its position as a boundary point in its phase space representation. For example, if ray 1 at the boundary wants to become a ray that occupies a position in phase space located internally, the following must happen: ray 1 would need to select one of its neighboring points to exchange positions with. However, the Liouville theorem requires the phase space volume occupied by all the rays of the system to have no source or sink. In other words, there is no generation, vanishing, or merging of the little squares representing each ray.

In order for white squares to interchange their positions in phase space at one point, they must be compressed, as in fig. 6.10, and eventually merged (the black square in fig. 6.11). However, merging the area of the phase space representations of any two rays will result in these rays having the same direction and position (black square), which renders them indistinguishable from then on. This makes it clear that two separate rays in a system cannot during their propagation within the optical system acquire both the same position and same direction, or occupy the same phase space area. This proves that

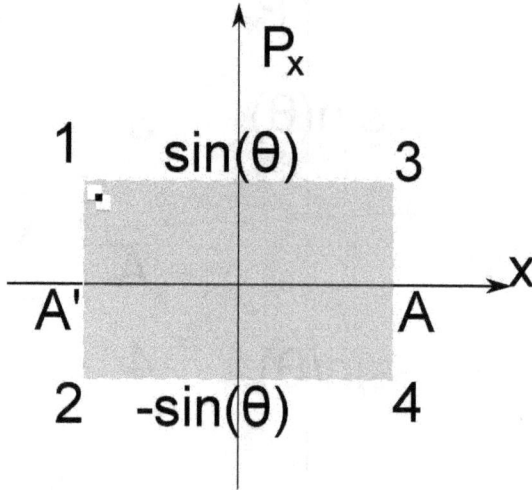

FIGURE 6.10: In order for ray 1 to interchange its phase space position, it will have to merge with a small phase space that was originally not on the boundary. This violates the Liouville theorem

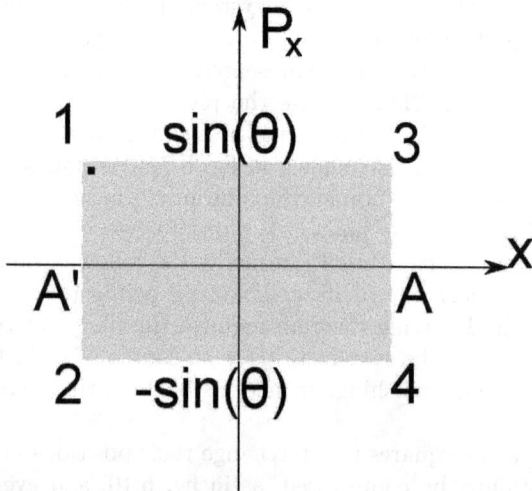

FIGURE 6.11: Merging of the edge ray with an internal ray in their phase space representations. Similarly, it violates the Liouville theorem.

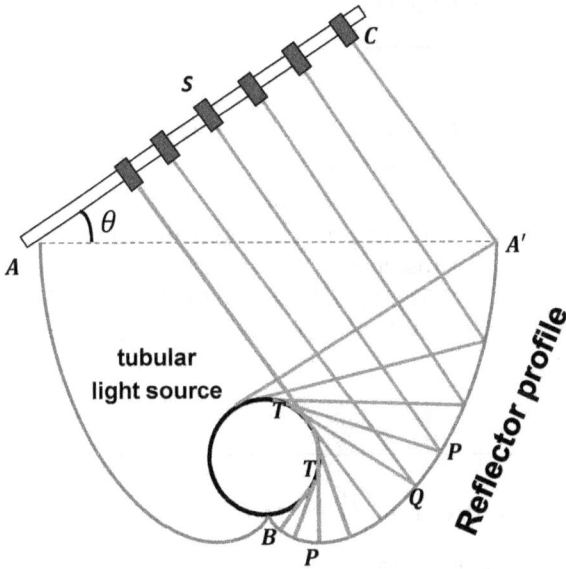

FIGURE 6.12: CPC design for a tubular absorber

the edge rays cannot interchange their positions with internal rays; they are stuck as edge rays. Any edge rays from the absorber must always be mapped to the edge rays of the radiation source, and vice versa.

Another example involves a convex absorber, such as the tubular absorber shown in fig. 6.12. To design for tilted edge rays at an angle of θ, we use the principle of redirecting the edge ray to the absorber tangent. We require that the absorber can only see within the designed angle θ, and we design reversely, evaluating the rays that come from the absorber. Starting from the bottom of the absorber B, we require that the unwinding string TP be the same length as the absorber section length $\overset{\frown}{BT}$, effectively forming an involute part of the concentrator $\overset{\frown}{BQ}$. In this part, the involute only redirects the rays from the absorber based on a 1:1 concentration ratio. Past this involute part we require the tangent ray TP of the absorber to be reflected into the extreme ray direction, forming the curve $\overset{\frown}{QA'}$, until the tangent of the reflector becomes vertical, resulting in the concentrator being at its maximum aperture. Another way of thinking of this design is to use the string method. We assign the point S to be on a slider which can move without friction, and it has a string attached to it. By tying this string to point P and around the absorber shape to point B, we require that the string $BTPS$ be always taut and have the same length. Now as we trace the point P along the curve, the point S slides freely, causing the string to be always perpendicular to the slider bar. This method also guarantees the constant optical path length between the edge-rays and the tangent points of the absorber.

An alternative way of designing the nonimaging concentrator is by replacing the idea of constant string length with correct ray angles. One starts from an arbitrary point at the bottom of the absorber, such as B. The curve of the reflector starts by reflecting the tangent ray of the absorber to its original direction to form the involute section $\overgroup{BQ}$ (TP is the same as PT). Then as soon as the reflector tangent can reflect the absorber tangent ray without being obstructed by the absorber, the tangent of the reflector should be constructed to do so [24] (PS is perpendicular to the bar AC). A shape generated with such a procedure is effectively the same as that generated by the method previously described.

6.2   Irradiance Vector $\vec{D}$ in Phase Space

In this section we are going to discuss the role of irradiance vector $\vec{D}$ in the framework of phase space. In section 1.4 we introduced the geometrical vector flux $\vec{J}$ and we studied its relation with irradiance vector $\vec{D}$, eq. 1.16. We defined the geometrical vector flux from the étendue; now we can write the étendue as the flux of $\vec{J}$ through a surface:

$$U = \int_s \vec{J} \cdot d\vec{S}. \tag{6.11}$$

Equation 6.11 connects vector $\vec{J}$, and also $\vec{D}$, with the framework of phase space. We can see this for 2D systems; in this case the $\vec{J}$ can be written

$$\vec{J} = \left(\int dp_z, \int dp_x \right), \tag{6.12}$$

and phase space is represented by (x, p_x). Then we can say that the width in the p_x direction in phase space is the J_z component of vector $\vec{J}$ at point x. This provides us an important property of 2D phase space: the histogram of phase space is the irradiance pattern. In a sense, for each x value of phase space diagram the width of the p_x is equal to the density of irradiance. This is the same as the J_z component we described previously, which is proportional to the irradiance D_z at a plane $z =$ constant.

To illustrate this idea we can return to the example of a previous section, a strip source of 10 mm length which emits light rays in a half angle of 30 deg. Figure 6.13 shows the phase space representation at $z = 5\,mm$, with the bar diagram representing the histogram. Figure 6.14 shows the irradiance pattern of this strip source at $Z = 5\,mm$. Note the correlation between the histogram bar and the irradiance pattern.

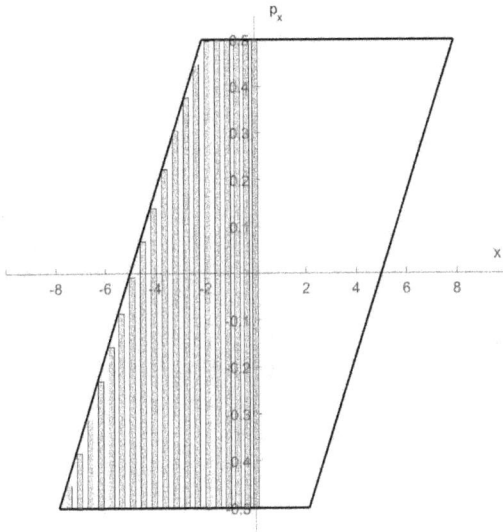

FIGURE 6.13: Phase space at $z = 5\,mm$ with bar diagram

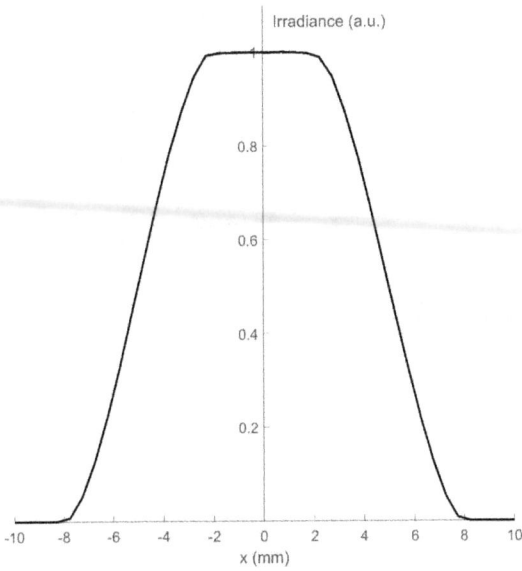

FIGURE 6.14: Irradiance pattern at $z = 5\,mm$

6.3   The $\frac{\theta_i}{\theta_o}$ Concentrator

The CPC concentrator studied in previous sections has an acceptance angle θ_i and an output angle of $\frac{\pi}{2}$. The $\frac{\theta_i}{\theta_o}$ concentrator [68] has an acceptance angle θ_i with a prescribed output angle θ_o. It consists of two reflectors PRQ and $P'R'Q'$ disposed symmetrically about the z axis, fig. 6.15. The section PR forms part of a parabola whose focus is at Q' and whose axis makes an angle θ_i with the z axis. RQ is a straight line section making an angle $\frac{1}{2}(\theta_o - \theta_i)$ with the z direction; no discontinuity in slope occurs at R. A meridional ray from the lower edge of the source, which is incident at R with an angle θ_i with the z axis, is reflected at an angle θ_o with the z axis. It will pass through the point Q' of the exit aperture, which is the intersection of this ray with a ray incident at point P with an incident angle θ_i with the z axis, fig. 6.15. It can be shown [68] that the geometry of this 2D arrangement is such that all rays incident on PP' within an angle $\pm\theta_i$ are collected on QQ' within an angle $\pm\theta_o$. From the concentration equation we can write

$$PP' \sin\theta_i = QQ' \sin\theta_o. \qquad (6.13)$$

It is possible to analyze the $\frac{\theta_i}{\theta_o}$ concentrator in the framework of $\vec{D}$ and phase space [69]. The transformation in phase space is a compression in x and an expansion in p_x, fig. 6.16. At plane PP' the maximum value of $|p_x|$ is $\sin\theta_i$; this maximum value increases smoothly, as z increases, up to $\sin\theta_o$ at plane PP'. Thus the limiting phase space contour at the exit is a rectangular area of width $2\sin\theta_o$ and height QQ', and from eq. 6.13 the phase space area enclosed at entrance and exit is equal, which is consistent with Liouville's Theorem. The phase space density is constant.

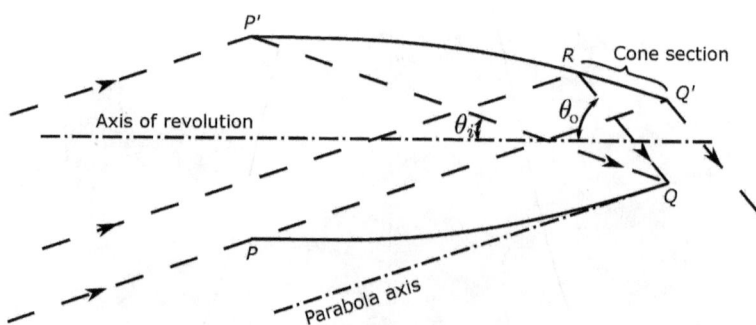

FIGURE 6.15: The $\frac{\theta_i}{\theta_o}$ concentrator, with $\theta_i = 18°$ and $\theta_o = 50°$

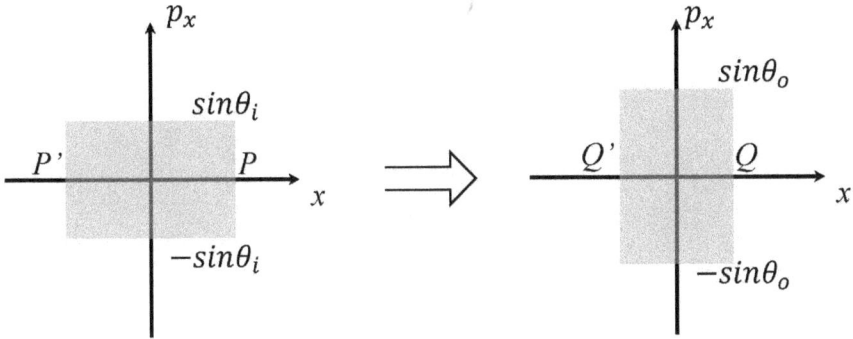

FIGURE 6.16: Entrance and exit phase space for the $\frac{\theta_i}{\theta_o}$ concentrator

Barnett [69] studied the phase space representation in planes inside the $\frac{\theta_i}{\theta_o}$ concentrator and also the irradiance pattern produced at those planes.

6.4 Cone Concentrator as Ideal 3D Phase Space Transfer Device

In previous sections in this chapter we defined the étendue as the product of the area of the source and the solid angle subtended by the radiation, and we also stated that this magnitude is conserved in free propagation or in perfect reflection or refraction optical systems. Let consider a spherical Lambertian source with radius r fig. 6.17; the field lines are radial from the center of the sphere, and it is possible to build a flux tube with the shape of a cone. Then the étendue will be conserved in the free propagation of radiation through the flux tube. We consider a portion of the source of area A_1: as a Lambertian source the étendue of this portion of the source is

$$U_1 = \Omega r^2 \sin^2\left(\frac{\pi}{2}\right) = \Omega r^2, \qquad (6.14)$$

where Ω is the solid angle subtended by the portion of the source to the center of the sphere. We can compute the étendue to the orthogonal surface of $\vec{D}$; in this case a spherical surface with radius R, considering that at point P the radiation cone has an angle θ, which is the same for any point of the orthogonal surface, by the geometry of the fig. 6.17

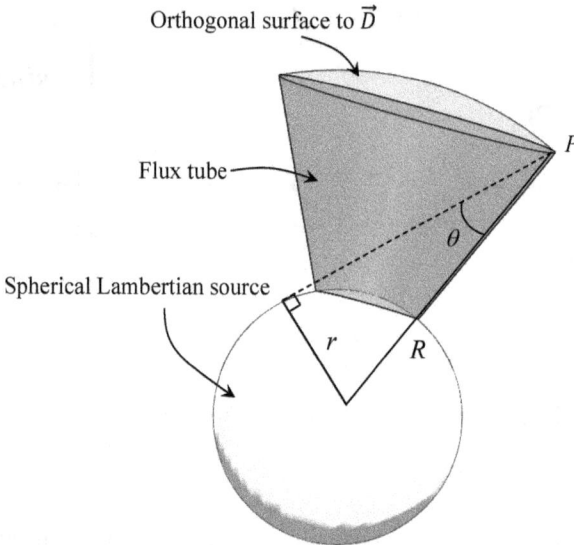

FIGURE 6.17: Conservation of étendue for spheric Lambertian surface

$$U_2 = \Omega R^2 \sin^2(\theta) = \Omega R^2 \frac{r^2}{R^2} = U_1, \qquad (6.15)$$

showing the conservation of étendue.

6.5   Fermi Proof of Phase Space Volume or Étendue Conservation

Consider the Legendre transformation of the optical Lagrangian and construct the Hamiltonian function

$$H = p_x \dot{x} + p_y \dot{y} - L \qquad (6.16)$$

and differentiate

$$dH = p_x d\dot{x} + p_y d\dot{y} + dp_x \dot{x} + dp_y \dot{y} - \left[\frac{\partial L}{\partial x} dx + \frac{\partial L}{\partial y} dy + \frac{\partial L}{\partial \dot{x}} d\dot{x} + \frac{\partial L}{\partial \dot{y}} d\dot{y} \right]. \quad (6.17)$$

Then

$$dH = -\dot{p}_x dx - -\dot{p}_y dy + \dot{x} dp_x + \dot{y} dp_y, \tag{6.18}$$

and this results in familiar Hamiltonian optics equations

$$\frac{\partial H}{\partial x} = -\dot{p}_x, \ \frac{\partial H}{\partial y} = -\dot{p}_y, \ \frac{\partial H}{\partial p_x} = \dot{x}, \ \frac{\partial H}{\partial p_y} = \dot{y}. \tag{6.19}$$

Notice that

$$\frac{\partial \dot{p}_x}{\partial p_x} = -\frac{\partial}{\partial p_x} \left(\frac{\partial H}{\partial x} \right) = -\frac{\partial}{\partial x} \left(\frac{\partial H}{\partial p_x} \right) = -\frac{\partial \dot{x}}{\partial x}, \text{ etc.} \tag{6.20}$$

which are Hamilton's equations of motion. Now we construct a vector $\vec{W} = (\dot{x}, \dot{y}, \dot{p}_x, \dot{p}_y)$; notice that

$$\nabla \vec{W} = \frac{\partial \dot{x}}{\partial x} + \frac{\partial \dot{p}_x}{\partial p_x} + \frac{\partial \dot{y}}{\partial y} + \frac{\partial \dot{p}_y}{\partial p_y} = 0. \tag{6.21}$$

This means the field of four-dimensional vector $\vec{W}$ has the important property of divergence being zero. In other words, the four-dimensional hyperspace of (x, y, p_x, p_y) has the property of conservation of volume as all the light rays evolve in an optical system.

Here we will also offer an intuitive analogy using an incompressible fluid, fig. 6.18. Let V be the volume surrounded by a closed surface S, where $\vec{v}$

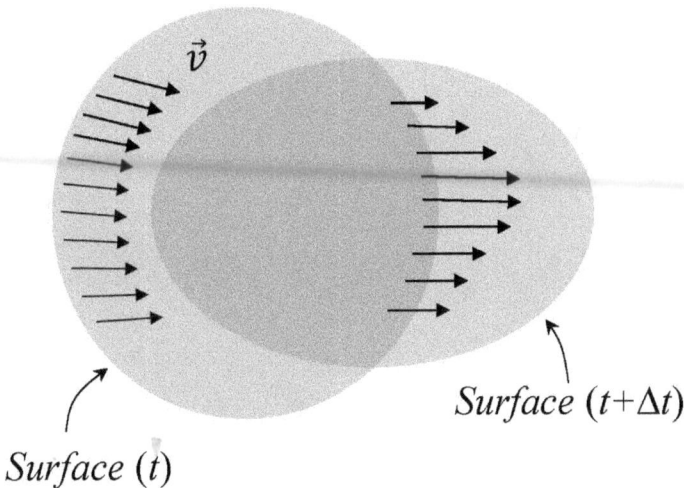

FIGURE 6.18: Phase space volume remains constant

is the velocity field of the small elements within . As the fluid starts to flow according the change of t, and the surface $S(t)$ starts to evolve into the surface $S(t + \Delta t)$, the enclosed volume of $V(t + \Delta t)$ will also change as

$$V(t + \Delta t) - V(t) = \oint_S \vec{v}\Delta t \cdot d\vec{S}, \qquad (6.22)$$

where $d\vec{S}$ is the surface vector pointing outward along the normal direction. Using Gauss's theorem

$$\oint_S \vec{v}\Delta t \cdot d\vec{S} = \int_V \nabla \cdot \vec{v}dV, \qquad (6.23)$$

where dV is the volume element. If $\nabla \cdot \vec{v} = 0$ everywhere, then obviously $V(t + \Delta t) - V(t) = 0$ or $V(t) =$constant. Applying this to the étendue conservation, we find that the four-dimensional volume (x, y, p_x, p_y) remains constant as it evolves with time. The vector field $\vec{W} = (\dot{x}, \dot{y}, \dot{p}_x, \dot{p}_y)$ replaces $\vec{v}$. This not only implies that the étendue is conserved, but also shows that there is no source or sink of field $\vec{W}$ as the light propagates.

A

The Edge Ray Theorem

A.1   Introduction

This theorem provides the key for the synthesis methods of concentrators and illuminators. It has been used as a powerful tool since the first designs of "nonimaging concentrators" in the 1960s, although its first proof was given by Miñano much later [70] [71]. Since then, other proofs and formulations have been proposed by Davis [72], Ries and Rabl [73].

For simplicity of explanation, we will consider here only the theorem in 2D geometry, partially following the approaches in Miñano [70]. The reader interested in the 3D treatment can consult Miñano [71]. First in section A.2, the theorem is formulated for the case of continuous inhomogeneous refractive index media. The theorem in the presence of sequential optical surfaces and nonsequential mirrors is considered in section A.3 and A.4 respectively. Finally in section A.5 the modifications to the theorem introduced by the presence of slope discontinuities in sequential optical surfaces is considered, and its application to an enjoyable example, due to Davis [72], that seems to constitute an offense against the edge-ray theorem at first sight.

A.2   The Continuous Case

Consider a certain two-parametric input bundle described at the plane $y = y_i$ by the points contained in the region M_i of the space $x - y - p$, and assume that we must design an optical system such that all the rays contained in the bundle M_i are cast into the exit bundle described at the plane $y = y_o$ by the region M_o, if reversed, also vice versa: all the rays of M_o if reversed, are cast into M_i. Consequently the trajectories $\Im(M_i)$ and $\Im(M_o)$ must be the same rod in $x - y - p$ space. Consider now the one-parametric bundle described by the boundary of region M_i (denoted by ∂M_i) and the one-parameter bundle described by ∂M_o. Since both bundles are one-parametric, $\Im(M_i)$ and $\Im(M_o)$ will be two surfaces. The theorem states that (under certain conditions that will be shown later) if $\Im(\partial M_i) = \Im(\partial M_o)$, them $\Im(M_i) = \Im(M_o)$. The inverse

DOI: 10.1201/9780367551605-A

theorem, also considered in the references, has no application for the design and will not be discussed here.

In order to prove this theorem, it is essential to consider the ray trajectories in the Hamiltonian formulation [24] as solutions of the following system of 1st order differential equations,

$$\frac{dx}{dt} = H_p \quad \frac{dp}{dt} = -H_x \quad \frac{dy}{dt} = H_q \quad \frac{dq}{dt} = -H_y \qquad (A.1)$$

where the Hamiltonian $H = p^2 + q^2 + r^2 - n^2(x, y, z)$ and t is a parameter along the ray. This parameter is $t = s/(2^{1/2}n) + k$, being s the arc length of the trajectory. The constant k can be set arbitrarily to zero. The solution must be consistent with $H = 0$.

Before proving the theorem, let us change the variables in the system of equations A.1. The new variables are x, y, θ, μ, where x and y are equal to the old Cartesian coordinates of a plane and θ and μ are related to p and q by the equations

$$p = n \cos \theta + \mu \quad q = n \sin \theta + \mu \qquad (A.2)$$

The proof of the theorem is clear using the new variables, and its conclusions can easily be extrapolated to the variables x, y, p, q. In these new variables, the defined regions are represented as shown in fig. A.1. The trajectories of the rays are the solutions of the system eq. A.1, consistent with $H = 0$, that is consistent with $n^2 = p^2 + q^2$. Using the new set of variables, the trajectories of the rays remain as the solutions of another system of equations involving x, y, θ, μ consistent with $\mu = 0$. This equation is necessarily a particular integral of the new system of equations, since $H = constant$ was a first integral of system (eq. A.1). Since $\mu = 0$, then the variable θ can be viewed as the angle formed by the ray and a line parallel to the x-axis.

Introducing the new variables in the system of eq. A.1 and taking into account that $\mu = 0$ the following system is obtained

$$\frac{dx}{dt} = n \cos \theta \quad \frac{dy}{dt} = n \sin \theta \quad \frac{d\theta}{dt} = n_x \sin \theta + n_y \cos \theta \qquad (A.3)$$

the first two equations are obtained directly. Equations $\frac{dp}{dt} = nn_x$ and $\frac{dq}{dt} = nn_y$ from system eq. A.1 lead to the same equations (the last one of system eq. A.3) when $\mu = 0$ and $\frac{d\mu}{dt} = 0$. Note that

$$\frac{dp}{dt} = \frac{dn}{dt} \cos \theta \frac{d\theta}{dt} + \frac{d\mu}{dt} \qquad (A.4)$$

and that

$$\frac{dn}{dt} = n_x \frac{dx}{dt} + n_y \frac{dy}{dt} = nn_x \cos \theta + nn_y \sin \theta \qquad (A.5)$$

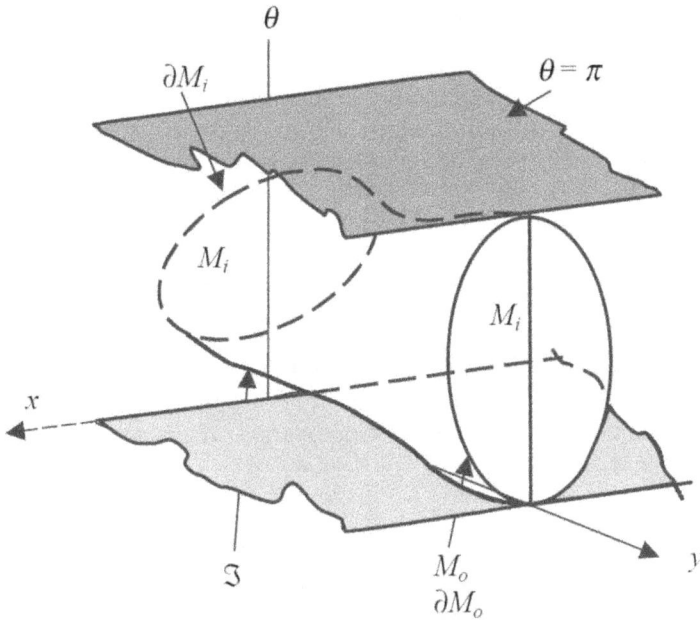

FIGURE A.1: The edge-ray theorem using coordinate $x - y - \theta$

Assume that the conditions of existence and uniqueness of the solutions of system of eq. A.1 are fulfilled. Then, there is only one solution crossing a given point (x_1, y_1, p_1, q_1) of this space $x - y - p - q$. Since the trajectories of the rays fulfill $n^2 = p^2 + q^2$, there are only two rays crossing a given point x_1, y_1, p_1 of this space $x - y - p$, one of them with $q_1 = (n^2 - p_1^2)^{1/2}$ and the other with $q_1 = -(n^2 - p_1^2)^{1/2}$, that is, one of them proceeds toward increasing y and the other toward decreasing y (excluding the case $q_1 = 0$). Now consider only the rays with $q_1 \geq 0$. In this case there is only one ray trajectory in the $x - y - p$ space crossing a given point x_1, y_1, p_1. The relationship between p and θ ($p = n\cos\theta$) establishes a point to point mapping of the space $x - y - p$ defined by $n^2 \geq p^2$. We assume that this mapping is continuous from U to U' and vice versa (for that n must be a continuous function), so it becomes a homeomorphism. If only rays with $q \geq 0$ are considered in $x - y - p$ space then any trajectory in U is transformed by the mapping into another trajectory in U', and vice versa, that is, the mapping relates the solutions of the system of eq. A.3 with those of the system of eq. A.1 together with $H = 0$.

We now give a proof of the theorem. Let M_i and M_o be two (simply connected) sets in the planes $y = y_i$ and $y = y_o$ (∂M_i and ∂M_o are the boundaries of these sets). Let $\Re$ be an open set in the $-y$ plane, including the trajectories of the rays of the bundle ∂M_i (or ∂M_o as we have assumed that the concentrator is built so that $\Im(\partial M_i) = \Im(\partial M_o)$), and let U be the region of the $x - y - \theta$ space such that (x, y, θ) belongs to U if (x, y) belongs to $\Re$. The hypotheses of the theorem are as follows:

1. The conditions for existence and uniqueness of the solution of the system of eq. A.3 are fulfilled in U, as are the conditions for the continuity of these solutions with respect to the initial conditions x_1, y_1, θ_1.

2. $\Im(\partial M_i) = \Im(\partial M_o)$; this surface will be called $\Im$ for simplicity.

3. $\Im$ is contained in the region $0 \leq \theta \leq \pi$.

4. $\Im$ is bounded.

From the first hypothesis, and particularly from the uniqueness of the solutions, it is derived that the trajectories of two rays, in the space $x - y - \theta$ cannot be common points. The equation of the surface $\Im$ can be written in parametric form as

$$
\begin{aligned}
x &= x(x_1(\tau), y_i, \theta_1(\tau), t) \\
y &= y(x_1(\tau), y_i, \theta_1(\tau), t) \\
\theta &= \theta(x_1(\tau), y_i, \theta_1(\tau), t)
\end{aligned}
\tag{A.6}
$$

where the parameters are t and τ. The functions $x_1(\tau)$ and $\theta_1(\tau)$ given the values of x_1 and θ_1 for the points of ∂M_i, so these functions are periodical, and give the same values when τ is increased by one period. The parameter t is the one appearing in eq. A.3. Without loss of generality, we can set $t = 0$ when a ray departs from ∂M_i to the value corresponding to a point of ∂M_o. The functions $x(x_1, y_1, \theta_1, t)$, $y(x_1, y_1, \theta_1, t)$ and $\theta(x_1, y_1, \theta_1, t)$ are the solutions of eq. A.3 and x_1, y_1, θ_1 are the initial conditions. Note that these functions are continuous with respect to all their parameters. By choosing the appropriate parameter τ a homeomorphism can be established between the surface $\Im$ and a tube where ∂M_e and ∂M_r are the boundaries of $\Im$. So we can conclude that $\Im$ together with M_e and M_r encloses a certain region of the $x - y - \theta$ space that will be called B.

Consider a ray described by the point p_i that belongs to the interior points of $M_i(\text{Int}(M_i))$. Four trajectories of this ray in the $x - y - \theta$ space can be considered, fig. A.2, as follows:

1. The trajectory intercepts the surface $\Im$.

2. It intercepts the surface M_i and leaves B through it.

3. The trajectory never leaves B.

4. The trajectory intercepts a point of $\text{Int}(M_o)$.

It is obvious that the trajectory departs from p_i. Option (1) is not possible because $\Im$ represents trajectories of rays, and these cannot cross in U. Option (2) is also impossible; the value of θ at the point p_i fulfills necessarily $0 < \theta < \pi$

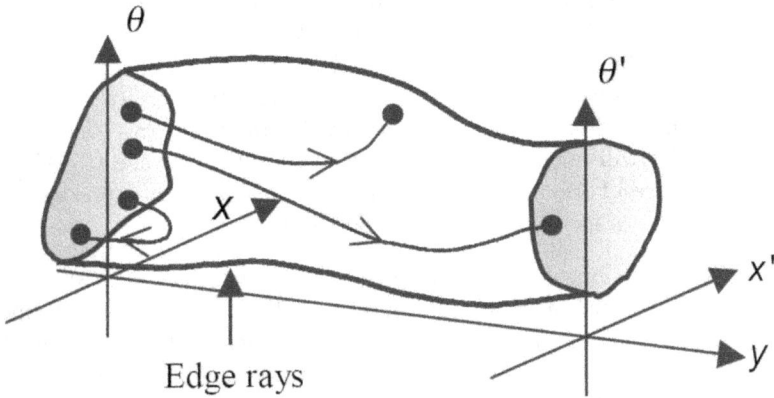

FIGURE A.2: Trajectories considered in the proof of the edge ray theorem

(note that if θ where zero or π for a point of $\text{Int}(M_i)$, due to the definition of the interior of a set, then the value of θ for some point of ∂M_i, which belongs to $\Im$, would not fulfill hypothesis $(\Im)$) and so the trajectory of the ray cannot be tangent to the plane $y = y_i$. Then this trajectory enters into B. If the trajectory comes back to M_i, it is necessary that θ takes, at some point of the trajectory, values zero or π (note that θ is continuous along the trajectory); that is, it is necessary that the trajectory intercepts the planes $\theta = 0$ or $\theta = \pi$ before reaches M_i, and these planes are outside B or coincide at some point with $\Im$. Then the trajectory must exit B before reaching M_i, which is contradictory with (2). Assume that the trajectory of the ray p_i never exits B, in other words, the trajectory is inside B for $t \in (0, \infty)$, or in terms of the arc length along the trajectory $s = \frac{t}{2^{1/2}n}$

$$\lim_{s \to \infty} y(s) = \int_{s=0}^{s=\infty} \sin\theta ds < y_o \tag{A.7}$$

Since $0 < \theta(s) < \pi$, $0 < \sin\theta(s) < 1$, and thus, eq. implies that

$$\lim_{s \to \infty} \sin\theta(s) = 0 \Rightarrow \lim_{s \to \infty} \theta(s) = 0 \text{ or } \pi \tag{A.8}$$

but then

$$\lim_{s \to \infty} x(s) = \int_{s=0}^{s=\infty} \cos\theta ds = \begin{cases} +\infty \text{ if } & \theta(\infty) = 0 \\ -\infty \text{ if } & \theta(\infty) = \pi \end{cases} \tag{A.9}$$

which contradicts that $\Im$ is bounded. Summarizing these last results, the only possibility for the trajectory of p_i is to intercept a point of M_o and so $\Im(\text{Int}(M_i))$ belongs to $\Im(M_o)$. A similar reasoning with respect to a ray

p_o of $\text{Int}(M_o)$ establishes that $\Im(\text{Int}(M_i))$, and so $\Im(\text{Int}(M_i)) = \Im(\text{Int}(M_o))$. Since $\Im(\partial M_i) = \Im(\partial M_o)$ by hypothesis, then $\Im(M_i) = \Im(M_o)$.

The homeomorphism between the region U of the space $x - y - \theta$ and the region U' of the space $x - y - p$ allows us to establish easily the edge-ray theorem in the space $x - y - p$. Effectively, if a ray intersects (or doesn't) a surface in U, it will intersect or not the corresponding surface U'. Also note that all the rays of M_i and M_o have $q \geq 0$. The theorem expressed in the variables $x - y - p$ has the following hypotheses:

1. The conditions for the existence and uniqueness of the solutions of the system of equations A.3 are fulfilled in U (U' is formed by the points x, y, p, q such that x, y belongs to $\Re$) as well as the condition for the continuity of these solutions with respect to the initial conditions.

2. $\Im(\partial M_i) = \Im(\partial M_o)$; this surface is called $\Im$.

3. The rays of $\Im$ have $q \geq 0$ all along their trajectories, as do the rays of M_i and M_o at $y = y_i$ and at $y = y_o$ respectively.

4. $\Im$ is bounded.

The fulfillment of these assumptions implies, as before, that $\Im(M_i) = \Im(M_o)$. The conditions mentioned in hypothesis (1) are that the right-hand side of eq. A.3 are continuous with respect to x, y, θ, t (or x, y, p, q). These conditions can be substituted by requiring that n, their first and their second partial derivatives are continuous in $\Re$ (this last conditions are sufficient but not necessary). The application of the theorem of conservation of étendue implies that the areas of M_i and M_o must be the same. However, it seems that M_i and M_o can be chosen such that their areas are different. This is not true. By means of the Stokes theorem applied to conservation of étendue theorem, it can be concluded that the condition $\Im(\partial M_i) = \Im(\partial M_o)$ also implies that the areas of M_i and M_o must be equal.

This proof of the edge-ray theorem can be viewed to input and exit aperture shapes different than straight lines by using the coordinate system $i - j$ defined by the field lines of $\vec{D}$ and its orthogonal lines. The entry and exit aperture would be contained in two lines $j =$ constant ($j = j_i$ and $j = j_o$ respectively).

The conditions of existence, uniqueness and continuity of the solutions with respect to the initial conditions are now fulfilled if the right-hand side of the equations A.6 are continuous with respect to i, j, u, v, t and fulfill the Lipschitz condition with respect to i, j, u, v. Condition (3) must now be replaced by $v \geq 0$. This implies that the optical directions cosine of the rays with respect to the j lines is not negative. Practical concentrators use reflectors and discontinuous refractive index distributions, neither of which is considered in the theorem just mentioned. The following sections will deal with these cases.

A.3   The Sequential Surface

Once we've seen the continuous refractive media case, consider the discontinuous case in which sequential optical mirrors or refractive surfaces are used. Sequential surfaces are defined as those on which all rays of the transmitted simply connected bundle impinge once. First, consider the line R of the $x - y$ as a 2D refractive surface. As in the former section, assume first that all rays have $q \geq 0$ (i.e. $0 \leq \theta \leq \pi$). Without loss of generality, the origin of the parameter t can be set at the points of the surface R for all the rays.

Since the whole bundle is refracted at that surface, we can identify the bundle before and after such incidence, which we will denote as M^- and M^+, respectively. Consider the points of the space (x, y, θ) of the bundle just before the refraction at every point $A = (x, y)$ of incidence on the surface R, which will be denoted as $x(t = 0^-)$, $y(t = 0^-)$ and $x(\theta = 0^-)$, and also the points of the space (x, y, θ) of the bundle just after the refraction $x(t = 0^+)$, $y(t = 0^+)$ and $x(\theta = 0^+)$.

The edge-ray theorem for the continuous media case of the former section applies for $t < 0$. Then the proof of the theorem for this discontinuous case is reduced to proof that if $\Im(\partial M^-) = \Im(\partial M^+)$, then $\Im(M^-) = \Im(M^+)$. In general the ray trajectories in the phase space become discontinuous at every point A (see fig. A.3). It will be sufficient to prove that the mapping of the points of M^- and M^+ is a homeomorphism. This is immediate because the mapping between M^- and M^+ is given by

$$x(0^+) = x(0^-)$$
$$y(0^+) = y(0^-)$$
$$\theta(0^+) = \theta_A + \sin^{-1}\left(\frac{n_A(0^-)}{n_A(0^+)}\sin(\theta(0^-) - \theta_A)\right)$$

(A.10)

where n_A is the discontinuous refractive index at A, and θ_A is the angle of the normal vector to R at A with respect to the x axis. The first two equations in eq. A.10 indicate that the ray trajectory in the $x - y$ space is continuous. If the surface is refractive, the third equation is the Snell's law at A. Since the hypothesis $0 \leq \theta \leq \pi$, the third equation is a homeomorphism.

The proof for the case of a reflective surface is analogous (in fact, setting $n_A^+ = n_A^-$ we obtain the reflection law). For the analogous formulation, all rays after the reflection must have $0 \leq \theta \leq \pi$ or $\pi \leq \theta \leq 2\pi$. Nevertheles, as in continuous case, the generalization for both refractive and reflective sequential surfaces can easily be extended to other input and exit aperture shapes by using the coordinate system $i - j$ defined by the field lines of $\vec{D}$ and the orthogonal lines.

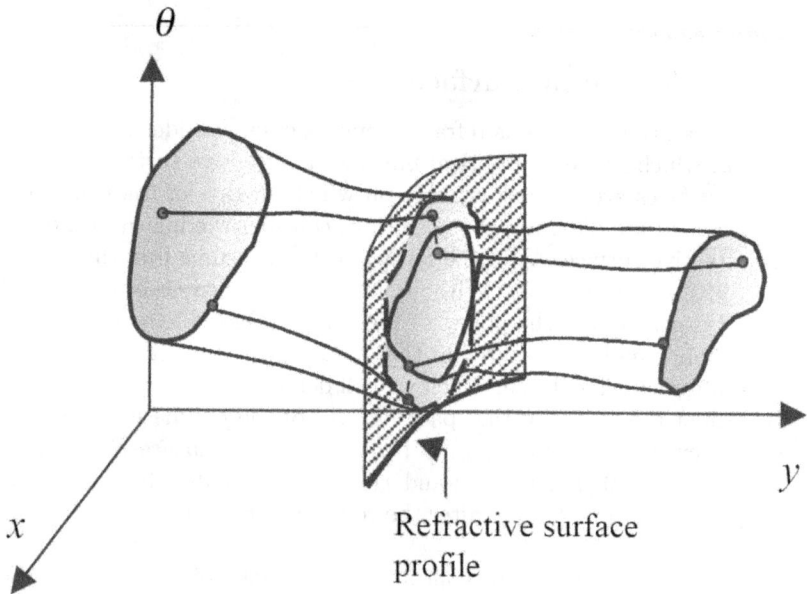

FIGURE A.3: The ray trajectories in the (x, y, θ) are discontinuous at the cylinder whose base is a sequential refractive surface

A.4   The Flowline Mirror Case

As seen above, the edge-ray theorem states that

$$\Im(\partial M_i) = \Im(\partial M_o) \Rightarrow \Im(M_i) = \Im(M_o) \tag{A.11}$$

In order to include nonsequential mirrors we must define other bundles of edge rays and reformulate the theorem. Let us consider nonsequential mirrors in which the bundle M that is reflected on the mirror is also presented. This mirror fulfills the following:

1. The rays of M that pass through the point A are edge rays and form a connected bundle at A. The angle formed between the tangent to the mirror at A and the ray r_a is smaller than that which it forms with r'_a. The rays of M at A situated between r_a and its symmetric with respect to the tangent are the edge rays ∂M_A, which do not extend beyond A (that is, they are edge rays only before reaching A).

2. The rays of M that pass through the point B are edge rays and form a connected bundle B. The angle formed between the tangent to the mirror at B and the ray r_b is smaller than that which it forms

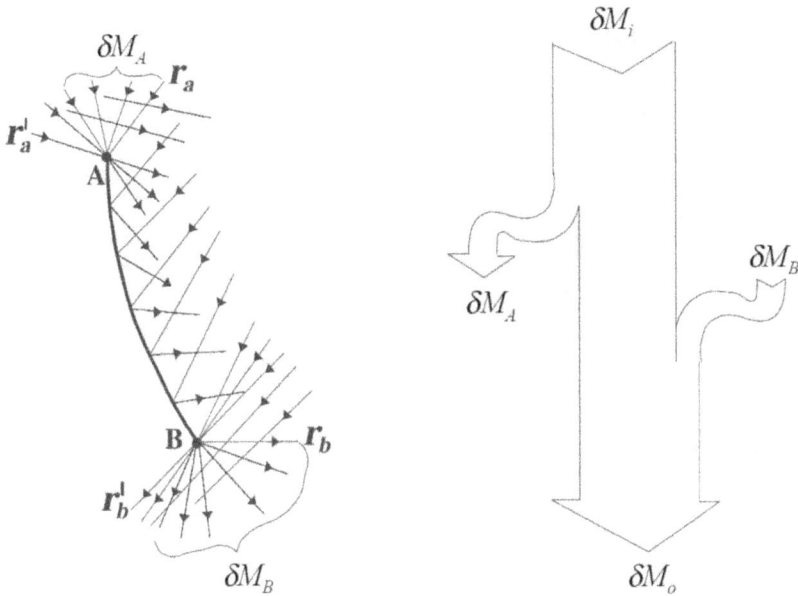

FIGURE A.4: Edge rays of the bundle reflected in a flowline mirror

with r'_b. The rays of M at B situated between r_b and its symmetric with respect to the tangent are the edge rays ∂M_B, which begin at B (that is, they are edge rays only after passing through B).

3. All edge rays impinging between A and B undergo a single reflection on the mirror.

In the case of fig. A.4a no more than one edge ray impinges at each point of the mirror between A and B. This is not a condition for the formulation of the theorem (although in practice it may be desirable).

If a concentrator is composed of sequential optical surfaces and a nonsequential mirror of the type just described, the edge ray theorem can be expressed as

$$\left.\begin{array}{c} \Im(\partial M_A) \cap \Im(\partial M_B) = \emptyset \\ \Im(\partial M_i) \cup \Im(\partial M_B) = \Im(\partial M_o) \cup \Im(\partial M_A) \end{array}\right\} \Rightarrow \Im(M_i) = \Im(M_o) \quad (A.12)$$

Thus, the design of such a concentrator is achieved by coupling, first, the edge rays of $\Im(\partial M_A)$ with the edge rays of the entry bundle $\Im(\partial M_i)$, second, the edge rays of $\Im(\partial M_B)$ with the edge rays of the exit bundle $\Im(\partial M_o)$ and third the rest of the rays of $\Im(\partial M_A)$ with the rest of the rays of $\Im(\partial M_B)$. Figure A.4b shows an explanatory diagram of this assignation of edge rays.

The demonstration of the edge ray theorem can be obtained under the same hypothesis as the continuous case. Now the region B is bounded by the surface

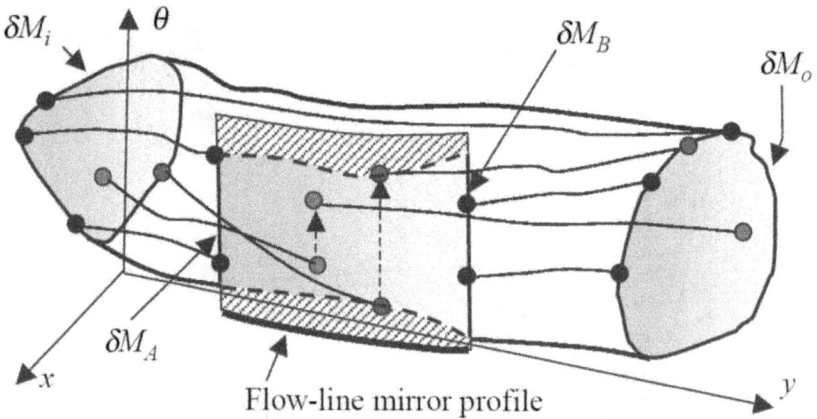

FIGURE A.5: Edge rays theorem in the presence of a flowline mirror. The ray trajectories in the (x, y, θ) are discontinuous at the cylinder whose base is flowline surface

of the edge ray trajectories and the cylinder obtained by the translation of the flowline mirror profile parallel to the θ axis, see fig. A.5. The edge rays ∂M_A and ∂M_B are represented as two vertical lines and belong to the boundary of B. The ray trajectories are discontinuous an the flowline mirror, suffering a vertical jump in the phase space (x, y, θ) as in the case of the sequential refraction of fig. A.5 given by the mapping.

Bibliography

[1] J.H. Lambert. *Photometria sive de mensura et gradibus luminis, colorum et umbrae*. Eberhardy Klett, 1760.

[2] J.R. Mehmke. Über die mathematische bestimmung der helligkeit in rämen mit tagesbeleuchtung, insbesondere gemäldesälen mit deckenlicht. *Zs. f. Math. u. Phys, 43*, page 41.

[3] V. Fock. Zur Berechnung der Beleuchtungsstärke, Zeitschrift für. *Physik, 28(1):* pages 102–113, 1925.

[4] A. Gershun. The light field. *Journal of Mathematics and Physics, 18*, pages 51–151, 1936.

[5] D.E. Spencer and P. Moon. *The Photic Field*. MIT press, Cambridge, 1981.

[6] R. Winston and W.T. Welford. *High Collection Nonimaging Optics*. Elsevier, 1989.

[7] Roland Winston and Walter T. Welford. Geometrical vector flux and some new nonimaging concentrators. *JOSA*, 69(4):532–536, 1979.

[8] A. Garcia-Botella, A.A. Alvarez-Balbuena, E. Bernabeu, D. Vazquez, A. Gonzalez-Cano. Hyperparabolic concentrator. *Applied Optics, 48(4):* 712–715, 2009.

[9] R. Winston, A. Garcia-Botella, L. Jiang. Flowline optical simulation to refractive/reflective 3D systems: optical path length correction. *Photonics, 101(6):* 1–13, 2019.

[10] Albert Einstein. The general theory of relativity. In *The Meaning of Relativity*, pages 54–75. Springer, 1922.

[11] Subrahmanyan Chandrasekhar. *Radiative transfer*. Courier Corporation, 2013.

[12] W.T. Welford and R. Winston. The ellipsoid paradox in thermodynamics. *Journal of Statistical Physics*, 28(3):603–606, 1982.

[13] Arthur Ashkin. Acceleration and trapping of particles by radiation pressure. *Physical Review Letters*, 24(4):156, 1970.

[14] R. Winston and X. Ning. Constructing a conserved flux from plane waves. *JOSA A*, 3(10):1629–1631, 1986.

[15] Enrico Fermi. *Notes on thermodynamics and statistics*. University of Chicago Press, 1966.

[16] R. Winston and W.T. Welford. Ideal flux concentrators as shapes that do not disturb the geometrical vector flux field: a new derivation of the compound parabolic concentrator. *JOSA*, 69(4):536–539, 1979.

[17] Ari Rabl. Edge-ray method for analysis of radiation transfer among specular reflectors. *Applied Optics*, 33(7):1248–1259, 1994.

[18] H. Hinterberger and R. Winston. Efficient light coupler for threshold čerenkov counters. *Review of Scientific Instruments*, 37(8):1094–1095, 1966.

[19] M. Ricketts R. Winston, L. Jiang. Nonimaging optics: a tutorial. *OSA Advances in Optics and Photonics, 10(2):* pages 484–511, 2018.

[20] J.R. Arvo. *Analytic methods for simulated light transport*. Thesis, Yale University, 1995.

[21] Parry Moon and Domina E Spencer. *Field theory handbook: including coordinate systems, differential equations and their solutions*. Springer, 2012.

[22] Valentin I. Ivanov and Michael K. Trubetskov. *Handbook of conformal mapping with computer-aided visualization*. CRC Press, 1994.

[23] Roland Winston. Light collection within the framework of geometrical optics. *JOSA*, 60(2):245–247, 1970.

[24] Roland Winston, Juan C. Miñano, Pablo G. Benitez, et al. *Nonimaging Optics*. Elsevier, 2005.

[25] Roland Winston. Cone collectors for finite sources. *Applied Optics*, 17(5):688–689, 1978.

[26] M.E. Barnett. The geometrical vector flux field within a compound elliptical concentrator. *Optik*, 54(5):429–432, 1980.

[27] Peretz Greenman. Geometrical vector flux sinks and ideal flux concentrators. *JOSA*, 71(6):777–779, 1981.

[28] Roland Schinzinger and Patricio A.A. Laura. *Conformal mapping: methods and applications*. Courier Corporation, 2012.

[29] Leonhard Euler. Principes généraux du mouvement des fluides. *Mémoires de l'académie des sciences de Berlin*, pages 274–315, 1757.

[30] Ulf Leonhardt. Optical conformal mapping. *Science*, 312(5781):1777–1780, 2006.

[31] Luis A. Alemán-Castañeda and Miguel A. Alonso. Study of reflectors for illumination via conformal maps. *Optics Letters*, 44(15):3809–3812, 2019.

[32] Ephraim M. Sparrow and Robert D. Cess. *Radiation Heat Transfer: Augmented Edition*. Routledge, 2018.

[33] Angel Garcia-Botella, Antonio Alvarez Fernandez-Balbuena, Daniel Vázquez, and Eusebio Bernabeu. Ideal 3D asymmetric concentrator. *Solar Energy*, 83(1):113–117, 2009.

[34] Joseph O'Gallagher, Roland Winston, and Walter T. Welford. Axially symmetric nonimaging flux concentrators with the maximum theoretical concentration ratio. *JOSA A*, 4(1):66–68, 1987.

[35] Lun Jiang and Roland Winston. Flow line asymmetric nonimaging concentrating optics. In *Nonimaging Optics: Efficient Design for Illumination and Solar Concentration XIII—Commemorating the 50th Anniversary of Nonimaging Optics*, volume 9955, page 99550I. International Society for Optics and Photonics, 2016.

[36] Manuel Gutiérrez, Juan C. Miñano, Carlos Vega, and Pablo Benítez. Application of lorentz geometry to nonimaging optics: new three-dimensional ideal concentrators. *JOSA A*, 13(3):532–540, 1996.

[37] George B. Arfken, Hans-Jurgen Weber, and Frank E. Harris. *Mathematical Methods for Physicists*. Waltham, MA. Elsevier, 2013.

[38] Angel Garcia-Botella, Antonio Alvarez Fernandez-Balbuena, and Eusebio Bernabeu. Elliptical concentrators. *Applied Optics*, 45(29):7622–7627, 2006.

[39] G.H. Derrick. A three-dimensional analogue of the hottel string construction for radiation transfer. *Optica Acta: International Journal of Optics*, 32(1):39–60, 1985.

[40] Lev Davydovič Landau, L.D. Landau, and E.M. Lifshitz. *The Classical Theory of Fields*, Volume 2. Butterworth-Heinemann, 1975.

[41] Wolfgang Pauli. Relativistic field theories of elementary particles. *Reviews of Modern Physics*, 13(3):203, 1941.

[42] John T. Winthrop. Propagation of structural information in optical wave fields. *JOSA*, 61(1):15–30, 1971.

[43] H.B. Phillips. Vector analysis, John Wiley & Sons. Inc., New York, 1933.

[44] Manuel Gutiérrez. Lorentz geometry technique in nonimaging optics. In *Conference Publications*, Volume 2003, page 386. American Institute of Mathematical Sciences, 2003.

[45] Barrett O'neill. *Semi-Riemannian geometry with applications to relativity*. Academic press, 1983.

[46] Gutiérrez Manuel García-Botella, Angel and Lun Jiang. Application of lorentz geometry to evaluation of irradiance patterns. In *Nonimaging Optics: Efficient Design for Illumination and Solar Concentration XVIII*, volume 12220, page 1222007. International Society for Optics and Photonics, 2022.

[47] Roland Winston. Dielectric compound parabolic concentrators. *Applied Optics*, 15(2):291–292, 1976.

[48] Pablo Gimenez-Benitez, Juan Carlos Miñano, José Blen, Rubén Mohedano Arroyo, Júlio Chaves, Oliver Dross, Maikel Hernández, and Waqidi Falicoff. Simultaneous multiple surface optical design method in three dimensions. *Optical Engineering*, 43(7):1489–1502, 2004.

[49] Angel García-Botella. Ideal flux field dielectric concentrators. *Applied Optics*, 50(28):5357–5360, 2011.

[50] Xiaohui Ning, Roland Winston, and Joseph O'Gallagher. Dielectric totally internally reflecting concentrators. *Applied Optics*, 26(2):300–305, 1987.

[51] Julio Chaves. *Introduction to nonimaging optics*. CRC press, 2008.

[52] William Ross McCluney. *Introduction to radiometry and photometry*. Artech House, 2014.

[53] John R. Howell, M. Pinar Mengüç, Kyle Daun, and Robert Siegel. *Thermal radiation heat transfer*. CRC press, 2020.

[54] Michael F. Modest and Sandip Mazumder. *Radiative heat transfer*. Academic press, 2021.

[55] Erich Merchand. *Gradient index optics*. Elsevier, 2012.

[56] Philip Gleckman, Joseph O'Gallagher, and Roland Winston. Concentration of sunlight to solar-surface levels using non-imaging optics. *Nature*, 339(6221):198–200, 1989.

[57] Lev Davidovich Landau and Evgenii Mikhailovich Lifshitz. *Statistical Physics*, Volume 5. Elsevier, 2013.

[58] Roland Winston, Chunhua Wang, and Weiya Zhang. Beating the optical liouville theorem: How does geometrical optics know the second law of thermodynamics? In *Nonimaging Optics: Efficient Design for Illumination and Solar Concentration VI*, volume 7423, page 742309. International Society for Optics and Photonics, 2009.

[59] G. Smestad, H. Ries, R. Winston, and E. Yablonovitch. The thermodynamic limits of light concentrators. *Solar Energy Materials*, 21(2-3):99–111, 1990.

[60] Lun Jiang and Roland Winston. Thermodynamic origin of nonimaging optics. *Journal of Photonics for Energy*, 6(4):047003, 2016.

[61] Harald Ries. Thermodynamic limitations of the concentration of electromagnetic radiation. *JOSA*, 72(3):380–385, 1982.

[62] Bennett K. Widyolar, Lun Jiang, and Roland Winston. Thermodynamics and the segmented compound parabolic concentrator. *Journal of Photonics for Energy*, 7(2):028002, 2017.

[63] Hoyt C. Hottel. Radiant heat transmission. W.H. McAdams, ed. Heat Transmission, 1954.

[64] Franz Mandl. *Statistical physics*, Volume 14. John Wiley & Sons, 1991.

[65] Max Planck. *The theory of heat radiation*. Blakiston, 1914.

[66] Tetsundo Sekiguchi and Kurt Bernardo Wolf. The hamiltonian formulation of optics. *American Journal of Physics*, 55(9):830–835, 1987.

[67] Herbert Goldstein, Charles P. Poole, and John Safko. *Classical mechanics*, Volume 2. Addison-Wesley Reading, MA, 1950.

[68] Ari Rabl and Roland Winston. Ideal concentrators for finite sources and restricted exit angles. *Applied Optics*, 15(11):2880–2883, 1976.

[69] M.E. Barnett. Optical-flow in an ideal light collector-the theta-i-theta-o concentrator. *Optik*, 57(3):391–400, 1980.

[70] Juan C. Miñano. Two-dimensional nonimaging concentrators with inhomogeneous media: a new look. *JOSA A*, 2(11):1826–1831, 1985.

[71] Juan C. Miñano. Design of three-dimensional nonimaging concentrators with inhomogeneous media. *JOSA A*, 3(9):1345–1353, 1986.

[72] P.A. Davies. Edge-ray principle of nonimaging optics. *JOSA A*, 11(4):1256–1259, 1994.

[73] Harald Ries and Ari Rabl. Edge-ray principle of nonimaging optics. *JOSA A*, 11(10):2627–2632, 1994.

Index

Absorptance, 134
Action-reaction principle, 137
Application of Lorentz geometry, 85
Associated signature, 83
Avogadro's number, 139

Blackbody radiator, 6, 127, 128, 133
Boltzmann's constant, 126

Cauchy-Riemann conditions, 38
Christoffel symbols, 88
Circular Lambertian source, 75, 120
Compound Elliptical Concentrator, 33, 37, 66
Compound Parabolic Concentrator, 10, 23, 35, 145
Concentration limit, 6, 28, 144
Cone concentrator, 155
Conformal mapping, 37
Conservation of energy, 14, 125
Conservative fields, 54
Continuous optical surface, 159
Contour integrals, 16, 53, 58, 60, 61, 76
Contour of the refracted cone, 103
Coordinate systems, 18, 35, 81
Cylindrical coordinates, 47

Dielectric concentrator, 93, 97
Differential geometry, 85
Discontinuous optical surface, 165

Edge ray principle, 9, 101
Edge rays cone, 10, 53, 57, 83, 85, 89, 101
Edge rays, curved cone, 122
Edge rays, elliptic cone, 86

Edge rays, reflected cone, 102, 111, 119
Edge rays, refracted cone, 101, 107, 115
Eigenvalues of G, 83, 89, 116, 119
Eigenvectors of G, 83, 89, 116, 119
Ellipsoidal Coordinates, 41
Elliptic disk, 90
Elliptical coordinates, 55
Emittance, 134
Entropy, 126
Equipotential surfaces, 78
Étendue, 6, 14, 126, 132, 142

Fermat's principle, 1, 66, 98, 102, 107, 111, 116, 119
Fermi proof of étendue conservation, 156
Field lines of truncated wedge, 24
Flowline asymmetric concentrator, 45
Flowline design method, 1, 16, 97
Flowline mirror, 166
Flux tube, 13, 14, 16, 96

Gauge transformation, 65, 70
Gauss's theorem, 14, 126
Gaussian beams, 20
Geodesics, 87
Geometrical Optics, 1, 4, 125, 128, 139
Geometrical vector flux, 6, 152
Gram matrix, 83

Hamiltonian optics, 142

Helmholtz decomposition theorem, 53

Histogram of phase space, 152

Hottel's strings, 130

Hyperboloid, 38

Hyperboloid, one-sheeted, 40

HyperElliptical Concentrator, 35, 37

Hyperparabolic Concentrator, 27, 35

Ideal source-receiver, 47

Inhomogeneous media, 121

Integrating factor, 80

Irradiance, 4

Irradiance distribution, 20, 70, 92, 107, 114, 117

Irrotational field, 53

Kirchhoff radiation law, 134

Lambertian plane sources, 61

Lambertian source, 5, 8, 38, 47, 54, 97, 128

LED, 21

Lightlike vectors, 83

Line integral, 15, 61

Liouville theorem, 149

Lorentz geometry, 53, 82, 114

Lorentzian scalar product, 82

Modulus of irradiance vector, 54, 84

Non-Lambertian source, 19

Normal component of $\vec{D}$, 16

Normal component of, $\vec{D}$, 95

Oblate Spheroidal Coordinates, 39

Optical interface, 93

Optical Lagrangian, 156

Optical momentum, 7, 139

Optical path length, 93

Optical path length difference, 93

Orthogonal refractive interfaces, 96

Phase space, 126, 139

Phase space volume, 141

Photometry, 1

Photon gas, 133

Planck's radiation law, 133

Plane convex lens, 107

Principle of reciprocity, 128

Quasipotential concentrator, 98

Quasipotential Fields, 54, 95

Quasipotential function, 80

Radiation pressure, 1, 6, 134

Radiative flux, 13, 61, 129

Radiative heat transfer, 3, 131

Radiometry, 1

Raytrace simulation, 2, 20, 56, 72, 75, 77, 92, 100, 107, 114, 117

Rectangular Lambertian source, 77

Reflective optical component, 93

Refracted irradiance vector, 95

Refractive media, 93

Rotational symmetry, 72, 88

Scalar potential, 15, 54

Second law of thermodynamics, 129

Snell's law, 95

Solar radiation, 6

Solenoidal field, 53, 93

Sphere Ellipse Paradox, 3

Spherical coordinate, 18

Spherical mirror, 111

Square Lambertian source, 77, 107, 111

Statistical physics, 126, 139

Stefan-Boltzmann law, 6, 125

Stoke's theorem, 58

Surface integral, 13, 61, 126

Tangential component of $\vec{D}$, 95

Temperature, 126

The $\frac{\theta_i}{\theta_o}$ concentrator, 154

Thermodynamic limit of concentration, 33

Thermoynamic variables, 125

Toroid, 47

Toroidal coordinates, 18, 47

Total internal reflection, 93
Transmission angle curves, 33, 100
Traslational symmetry, 70

Vector potential, 54, 63, 70, 76
Velocity field, 158
View factor, 12, 131

For Product Safety Concerns and Information please contact our EU
representative GPSR@taylorandfrancis.com
Taylor & Francis Verlag GmbH, Kaufingerstraße 24, 80331 München, Germany

www.ingramcontent.com/pod-product-compliance
Lightning Source LLC
Chambersburg PA
CBHW070723220326
41598CB00024BA/3276

* 9 7 8 0 3 6 7 5 5 1 6 3 6 *